Tourism and Earthquakes

ASPECTS OF TOURISM

Series Editors: **Chris Cooper**, *Leeds Beckett University, UK,* **C. Michael Hall**, *University of Canterbury, New Zealand* and **Dallen J. Timothy**, *Arizona State University, USA*

Aspects of Tourism is an innovative, multifaceted series, which comprises authoritative reference handbooks on global tourism regions, research volumes, texts and monographs. It is designed to provide readers with the latest thinking on tourism worldwide and in so doing will push back the frontiers of tourism knowledge. The series also introduces a new generation of international tourism authors writing on leading edge topics.

The volumes are authoritative, readable and user-friendly, providing accessible sources for further research. Books in the series are commissioned to probe the relationship between tourism and cognate subject areas such as strategy, development, retailing, sport and environmental studies. The publisher and series editors welcome proposals from writers with projects on the above topics.

All books in this series are externally peer-reviewed.

Full details of all the books in this series and of all our other publications can be found on http://www.channelviewpublications.com, or by writing to Channel View Publications, St Nicholas House, 31-34 High Street, Bristol BS1 2AW, UK.

ASPECTS OF TOURISM: 90

Tourism and Earthquakes

Edited by
C. Michael Hall and Girish Prayag

CHANNEL VIEW PUBLICATIONS
Bristol • Blue Ridge Summit

DOI https://doi.org/10.21832/HALL7864
Names: Hall, C. Michael, editor. | Prayag, Girish, editor.
Title: Tourism and Earthquakes/Edited by C. Michael Hall and Girish Prayag.
Description: Blue Ridge Summit: Channel View Publications, 2020. | Series: Aspects of Tourism: 90 | Includes bibliographical references and index. | Summary: 'This book examines the relationship between tourism and earthquakes through all stages of a disaster. It discusses the measures for managing tourism after earthquakes and examines the means to mitigate the impacts of earthquakes. It provides insights into the ethical, commercial and socioeconomic issues facing tourism after a major earthquake' – Provided by publisher.
Identifiers: LCCN 2020024657 (print) | LCCN 2020024658 (ebook) | ISBN 9781845417857 (Paperback) | ISBN 9781845417864 (Hardback) | ISBN 9781845417871 (PDF) | ISBN 9781845417888 (ePub) | ISBN 9781845417895 (Kindle Edition)
Subjects: LCSH: Tourism – Environmental aspects. | Earthquakes.
Classification: LCC G156.5.E58 T67 2020 (print) | LCC G156.5.E58 (ebook) | DDC 338.4/791 – dc23
LC record available at https://lccn.loc.gov/2020024657
LC ebook record available at https://lccn.loc.gov/2020024658
Library of Congress Cataloging in Publication Data
A catalog record for this book is available from the Library of Congress.

British Library Cataloguing in Publication Data
A catalogue entry for this book is available from the British Library.

ISBN-13: 978-1-84541-786-4 (hbk)
ISBN-13: 978-1-84541-785-7 (pbk)

Channel View Publications
UK: St Nicholas House, 31-34 High Street, Bristol, BS1 2AW, UK.
USA: NBN, Blue Ridge Summit, PA, USA.

Website: www.channelviewpublications.com
Twitter: Channel_View
Facebook: https://www.facebook.com/channelviewpublications
Blog: www.channelviewpublications.wordpress.com

Copyright © 2021 C. Michael Hall, Girish Prayag and the authors of individual chapters.

All rights reserved. No part of this work may be reproduced in any form or by any means without permission in writing from the publisher.

The policy of Multilingual Matters/Channel View Publications is to use papers that are natural, renewable and recyclable products, made from wood grown in sustainable forests. In the manufacturing process of our books, and to further support our policy, preference is given to printers that have FSC and PEFC Chain of Custody certification. The FSC and/or PEFC logos will appear on those books where full certification has been granted to the printer concerned.

Typeset by Riverside Publishing Solutions.

Contents

Tables, Figures, Plates	vii
Contributors	ix
Acknowledgements	xi

1 Earthquakes and Tourism: Impacts, Responses and Resilience – An Introduction 1
C. Michael Hall and Girish Prayag

2 The Resilience of a Tourist Destination: Seismic Risk Perception by Tourism Operators in the Etna Area, Italy 36
Barbara Martini and Marco Platania

3 Crisis Communication Systems and Earthquake Preparedness of Tourism Sectors in LDCs: A Study on Nepal 51
Subhajit Das and Premangshu Chakrabarty

4 Mitigating Earthquake and Tsunami Risks in Coastal Tourism Sites in Bali 65
I Nengah Subadra

5 It is Not Just About a Convention Centre: Expectations and Disillusions from Tourism-Relevant Stakeholders in Post-Earthquake Christchurch 82
Alberto Amore

6 Bringing Relief to a Natural Disaster Zone Through 'Being a Tourist': The Case of the 2010 Earthquake in Haiti 99
Nigel D. Morpeth

7 The Tourism Industry Response in Assisting Resident Evacuees after the 2016 Kumamoto Earthquakes 115
Atsuko Hashimoto and David J. Telfer

8 Handicraft Shopping Tourism after the Jogjakarta
   Earthquake: Recovery Network, Risk Perceptions and
   the Implications   138
   Andri N.R. Mardiah and Jon C. Lovett

9 Earthquakes, Psychological Resilience and
   Organizational Resilience: Tourism Entrepreneurs in
   Kaikōura, New Zealand   153
   Shupin (Echo) Fang, Girish Prayag and Lucie K. Ozanne

10 Ghost Towns and Tourism: L'Aquila, Italy Post-Earthquake   168
   Daniel Wright

11 Conclusion: Earthquakes and Tourism – An Emerging
   Research Agenda   193
   Girish Prayag and C. Michael Hall

   Index   203

Tables, Figures, Plates

Tables

1.1	Top 10 costliest world earthquakes and tsunamis by insured losses, 1980–2001	2
1.2	Major focus of chapters in relation to stages of disaster management cycle	26
2.1	Socioeconomic characteristics of respondents	41
2.2	Provisional timeframes of a fault earthquake in the Etna area	42
2.3	Level of threat due to earthquake	42
2.4	Earthquake risk: Sources of information	44
2.5	Community preparedness initiatives	45
2.6	Respondent perception of preparedness with respect to a seismic event	45
2.7	Perceptions of community belonging	46
2.8	Preparation acts for business resilience	46
4.1	Number of foreign visitors to Indonesia and Bali, 1969–2018	67
4.2	Number of rooms in classified hotels, non-classified hotels and other accommodation in the southern coastal region of Bali	69
7.1	Kumamoto Earthquake (earthquakes recorded stronger than M 5.0) between 14 April and 31 August 2016	118
7.2	Earthquake frequency by magnitudes (14 April 2016–30 June 2017)	119
8.1	Core business of respondents	142
8.2	The recovery period of business entities after the Jogjakarta Earthquake	144
10.1	Ghost tourism cities	174
10.2	Research participants: Local community members	179
10.3	Research participants: Tourists	179

Figures

1.1	System dimensions of tourism in earthquake affected destinations	11
2.1	Fault earthquake risk: Perception of preparedness	43
2.2	Discussion on earthquake preparedness	47
2.3	Opinion regarding consequences of fault earthquake	47
2.4	Opinion regarding consequences of fault earthquake	48
7.1	Kumamoto Earthquake epicentres and surrounding area	116
8.1	The most widely perceived business difficulties after the earthquake	145

Plates

4.1	Hotel displaying tsunami evacuation signage at Sanur Beach, 2017	73
4.2	Temporary Tsunami Evacuation Centre in Serangan Village	75
7.1	Aso Farmland accommodation poster at the entrance to Aso Farmland complex	124

Contributors

Alberto Amore, School of Business, Law and Communications, Southampton Solent University, UK.

Premangshu Chakrabarty, Department of Geography, Visva-Bharati, India.

Subhajit Das, Department of Geography, Presidency University, India.

C. Michael Hall, Department of Management, Marketing and Entrepreneurship, University of Canterbury, New Zealand; Geography Research Unit, University of Oulu, Finland; School of Business and Economics, Linnaeus University, Kalmar, Sweden; Department of Service Management and Service Studies, Lund University, Helsingborg, Sweden.

Atsuko Hashimoto, Department of Geography and Tourism Studies, Brock University, Canada.

Jon C. Lovett, School of Geography, University of Leeds, UK.

Andri N.R. Mardiah, School of Geography, University of Leeds, UK; The Ministry of National Development Planning (BAPPENAS), Indonesia.

Barbara Martini, Department of Economics and Finance, University of Rome, Italy.

Nigel D. Morpeth, Culture Place Policy Institute, University of Hull, UK.

Lucie Ozanne, Department of Management, Marketing and Entrepreneurship, University of Canterbury, New Zealand.

Marco Platania, Department of Educational Sciences, University of Catania, Italy.

Girish Prayag, Department of Management, Marketing and Entrepreneurship, University of Canterbury, New Zealand.

I Nengah Subadra, Sekolah Tinggi Pariwisata Triatma Jaya / Tourism Institute of Triatma Jaya, Jalan Kubu Gunung, Banjar Tegal Jaya, Kelurahan Dalung, Kecamatan Kuta Utara, Kabupaten, Bali, Indonesia.

David J. Telfer, Department of Geography and Tourism Studies, Brock University, Canada.

Daniel Wright, University of Central Lancashire, UK.

Shupin (Echo) Fang, Department of Management, Marketing and Entrepreneurship, University of Canterbury, New Zealand.

Acknowledgements

Writing a book on the relationships between earthquakes and tourism after having experienced their impacts directly – in the cases of the Christchurch earthquake sequence and the Kaikoura earthquakes – is a strange experience. On the one hand there is an attempt to retain a sense of distance and detachment which is integral to much academic writing and research. Indeed, having that personal perspective arguably makes one's understanding of the relationships between earthquakes and their effects on destinations and the wider tourism system all the stronger. On the other, it is extremely personal, arousing both memories of the effects of the earthquake events but also one's direct feelings concerning recovery and rebuild, the actions of government and political and economic interests and the way in which lives and places have been interrupted and set on new trajectories, not all of which are for the better. As we have recognised elsewhere, these are feelings that arguably only researchers who have been through such events understand. Nevertheless, they provide an important reality for conducting research on disasters and tourism, and earthquakes in particular.

Having lived and worked through a number of disasters ranging from the Christchurch earthquake sequence to Brexit, Michael would like to thank a number of colleagues with whom he has undertaken related conversations and research on a range of disasters over the years. In particular, thanks go to Bailey Adie, Alberto Amore, Dorothee Bohn Chris Chen, Tim Coles, Hervé Corvellec, David Duval, Martin Gren, Stefan Gössling, Johan Hultman, Dieter Müller, Paul Peeters, Yael Ram, Anna Laura Raschke, Jarkko Saarinen, Dan Scott, Anna Dóra Sæþórsdóttir, Allan Williams, Kimberley Wood and Maria José Zapata-Campos for their thoughts, as well as for the stimulation of Agnes Obel, Ann Brun, Beirut, Paul Buchanan, Nick Cave, Bruce Cockburn, Elvis Costello, Stephen Cummings, David Bowie, Ebba Fosberg, Mark Hollis, Margaret Glaspy, Aimee Mann, Larkin Poe, Vinnie Reilly, Henry Rollins, Matthew Sweet, Henry Wagon and The Guardian, BBC6, JJ, and KCRW – for making the world much less confining. Special mention must also be given to the Malmö Saluhall; Balck, Packhus and Postgarten in Kalmar; and Nicole Aignier and the Hotel Grüner Baum in Merzhausen. Finally, and most importantly, Michael would like to thank the Js and the Cs who stay at home and mind the farm.

Girish would like to thank a number of colleagues with whom he has undertaken disaster management research in the past few years for sharing their knowledge and insights. In particular, thanks go to Caroline Orchiston, Mesbahuddin Chowdhury, Lucie Ozanne, Sam Spector, Peter Fieger, Alberto Amore, Deborah Blackman, Hitomi Nakanishi, Ben Freyens, Joerg Finsterwalder, Alistair Tombs, Chris Chen and Sussie Morrish. Also, Girish would like to thank family and friends who have supported him over the years.

We also wish to gratefully acknowledge the help and support of Jody Cowper for proofreading and editing. Finally, we would both like to thank all at Channel View for their continuing support.

1 Earthquakes and Tourism: Impacts, Responses and Resilience – An Introduction

C. Michael Hall and Girish Prayag

Introduction

Earthquakes are a form of natural disaster with substantial human, economic and environmental effects. As many as 500,000 earthquakes occur around the planet each year, of which only about 20% are strong enough to be felt, with approximately 100 causing significant amounts of damage (United States Geological Survey (USGS), 2019). Nevertheless, large earthquakes can cause substantial loss of human life, substantially affect housing and infrastructure, and have major economic impacts. Large, damaging earthquakes (magnitude 5.5 or greater) are relatively rare in developed countries, averaging fewer than seven events per year since 1985 (Federal Reserve Bank of Kansas City, 2016). Nevertheless, their impacts can be substantial and are a function of a variety of factors including magnitude, duration, depth, landscape and geology, population density, construction practices and location of the epicentre.

Of the 48 earthquakes in developed countries from 1985 to 2015 of magnitude 5.5 or greater for which there are damage estimates the amount of economic damage varied significantly, from about $2 million to more than $232 billion (in 2015 dollars) (Table 1.1). Nevertheless, the median economic damage of earthquakes of magnitudes greater than 6.5 since 1985 ($628 million) was about 3.5 times higher than the median for earthquakes of magnitude 5.5 to 6.5 ($178 million), while the median damage of large earthquakes in areas with populations greater than 250,000 (nearly $2 billion) was nearly 75 times greater than for those in areas with populations below 250,000 ($28 million) (Federal Reserve Bank of Kansas City, 2016). In the United States the Federal Emergency Management Agency (FEMA), USGS and the Pacific Disaster Center (PDC) (2017) estimated that, in the United State, the annualized

Table 1.1 Top 10 costliest world earthquakes and tsunamis by insured losses, 1980–2001

Rank	Date	Location	Losses when occurred Overall (US$m)	Losses when occurred Insured (US$m)	% insured of losses (US$m)	Fatalities
1	March 11, 2011	Japan: Aomori, Chiba, Fukushima, Ibaraki, Iwate, Miyagi, Tochigi, Tokyo, Yamagata. Includes tsunami	210,000	40,000	19.04%	15,880
2	Feb. 22, 2011	New Zealand: Canterbury, Christchurch, Lyttelton	24,000	16,500	68.75%	185
3	Jan. 17, 1994	USA, California: Northridge, Los Angeles, San Fernando Valley, Ventura, Orange	44,000	15,300	34.77%	61
4	Feb. 27, 2010	Chile: Concepcion, Metropolitana, Rancagua, Talca, Temuco, Valparaiso. Includes tsunami.	30,000	8000	26.66%	520
5	Sep. 4, 2010	New Zealand: Canterbury, Christchurch, Timaru, Kaiapoi, Lyttelton	10,000	7400	74%	0
6	Apr. 14-16, 2016	Japan: Kumamoto, Aso, Chuo Ward, Mashiki, Minamiaso, Oita, Miyazaki, Fukuoka, Yamaguchi	32,000	6200	19.37%	205
7	Jan. 17, 1995	Japan: Hyogo, Kobe, Osaka, Kyoto	100,000	3000	30%	6430
8	Nov. 13, 2016	New Zealand: Canterbury, Kaikoura, Wellington, Marlborough, Picton	3900	2100	53.85%	2
9	Jun. 13, 2011	New Zealand: Canterbury, Christchurch, Lyttelton	2700	2100	77.78%	1
10	Sep. 19, 2017	Mexico: Puebla, Morelos, Greater Mexico City	6000	2000	33.33%	369

Source: After Insurance Information Institute, (2019)

earthquake loss (AEL) is $6.1 billion per year, a figure almost equal to the historic annual losses experienced from floods and hurricanes, while the potential exposure to earthquakes is of the order of approximately 59 trillion USD.

However, while the greatest economic losses from earthquakes are attributed to those in developed countries it is arguably developing countries that are proportionately worse affected and which also face major problems in allocating scarce resources in mitigating earthquake risk (Steckler et al., 2018). For example, when the relative values between nations based on a division of economic losses incurred at the time of the earthquake disaster as compared to GDP are considered then Armenia, Turkmenistan, Haiti, Nicaragua, Wallis and Futuna, North Macedonia and Chile have the highest relative ratios (Daniell et al., 2011). With respect to fatalities as a result of earthquakes China, Haiti, Indonesia, Iran, Japan and Turkmenistan have had the highest death and injury counts since 1900 in terms of absolute numbers, while Turkmenistan and Armenia have the highest relative fatality rates globally (Daniell et al., 2011).

Although the impacts of earthquakes and associated secondary disasters, such as tsunami and landslides, on tourism are substantial, the effects are often missed in the official figures. This is because while the loss of tourism specific infrastructure, such as hotels, may be covered the economic value of the expenditures of a temporary population such as tourists can be lost from official figures. This is especially because, being mobile, tourists can switch from one destination to another within the same national or even regional economy. Nevertheless, given its substantial direct and indirect contribution to economies on both developed and developing countries, tourism does matter.

This chapter introduces the phenomenon of earthquakes and their impacts on destinations, communities, businesses and individuals within the tourism system. The chapter provides a review of some of the major themes in research on earthquakes and tourism and also positions earthquakes and their impacts within the context of contemporary interest on tourism and resilience. The chapter concludes with an outline of the book.

Earthquakes, Disasters and Impacts

An earthquake is any shaking of the Earth's surface as a result of a sudden release of energy in the Earth's lithosphere that creates seismic waves. In its most general usage, the word earthquake therefore refers to any seismic event, whether natural, i.e. the result of a faults at the boundary or interior of tectonic plates, volcanic events, tidal forces or, more rarely, asteroid or meteorite impact; or caused by humans, i.e. fracking, nuclear or other large explosions. Earthquakes are happening around the Earth all the time, many of these are never felt by people. However, there are probably few natural events that can cause as much fear than feeling an earthquake and, arguably, some of the most damaging disasters are those that arise as the result of large earthquake events and the subsequent sequence of aftershock and other events that they can trigger such as tsunami, landslides and even volcanic activity. Some earthquake events and sequences can even come part of popular culture or at least personal and collective psychology, e.g. Pompei, the 1755 Lisbon earthquakes and tsunami and the 2004 Indian Ocean earthquake and tsunami. The 2011 Tōhoku earthquake in Japan, activated a tsunami with 30-foot waves and led to nearly 18,729 deaths and 2666 missing (Henderson, 2013; Fukui & Ohe, 2019), with footage of the tsunami being broadcast live around the world. Similarly, footage of the Indian Ocean Boxing Day (26 December) tsunami of 2004 that followed from the 9.1 (Mw) earthquake off the coast of the Indonesian Island of Sumatra that killed at least 225,000 people across several countries bordering the Indian Ocean including India, Indonesia, Sri Lanka and Thailand reminds us of the destructive nature of earthquakes and their impacts on humans, society and countries (Moeller, 2006; Mäntyniemi, 2012; Bonati, 2015).

Earthquake events and their impacts are clearly often devasting for the communities and people that are affected. From a tourism perspective this means members of the host community, visiting tourists, tourism businesses, destination infrastructure as well as destination and business image. Although it may seem trite to be discussing tourism in the context of earthquakes and disasters it needs to be remembered that tourism is economically important for many earthquake affected locations. If they lose tourism, they lose jobs and the economic capacity to rebuild, including damaged infrastructure, heritage and facilities that are used by the local population as well as visitors (Huan *et al.*, 2004; Mendoza *et al.*, 2012; Tang, 2014). To understand how tourism and tourists are affected by earthquakes is then essential to being able to build more resilient places, economies, businesses and destinations, and to be proactive with respect to being able to better help people when disaster does come (Huang & Min, 2002; Orchiston, 2013; Ghmire, 2016; Hall *et al.*, 2016).

The majority of earthquakes that people experience are fault earthquakes. Fault earthquakes happen when two earth blocks suddenly slip past one another, with the surface where the slip occurs known as the fault or fault plane (USGS, 2019). The main earthquake is called the mainshock and this is usually followed by a number of aftershocks, which may continue for weeks, months or even years after the main earthquake event or mainshock (USGS, 2019). Four different types of faults have been identified that explain the sudden 'jerky' feeling experienced during an earthquake. According to GNS Science,

New Zealand (2019), the four types of faults are: normal (move up and down), reverse (thrust), strike slip (move left and right or *vice versa*) and oblique slip (various combinations of the previously described movements). Along with the different types of fault earthquakes, volcanic related earthquakes are also significant. These are caused by the movement of magma beneath the Earth's surface which can lead to earthquake swarms as well as more violent earthquakes. Such earthquake activity is often an indicator of potential eruptions, as in the case of Mount St Helens (Foxworthy & Hill, 1982) or Mount Etna (Martini & Platania, this volume)

Unlike many other types of disasters, earthquakes have the capacity to generate other deadly disasters other than those arising from ground shaking, liquefaction, building collapse and falling masonry, including tsunamis and landslides, as well as those arising from their effects on chemical, oil and nuclear facilities, such as nuclear meltdown in Fukushima as a result of a tsunami (Hasegawa, 2012; Rangel & Lévêque, 2014). Very often there is no early warning which makes it impossible to precisely anticipate the location and intensity of the earthquake (Tsai & Chen, 2010). Typically earthquakes are treated as acute events with relatively short periods of impact and response that transform in the recovery phase (Becker *et al.*, 2019). However, this is somewhat misleading as not only may an earthquake sequence last for a considerable length of time, even years, as was the case of the Christchurch earthquakes in New Zealand. Similarly, the rebuild, insurance and psychological impacts of earthquakes can last for many years, as was also the case of the Christchurch earthquakes (Amore & Hall, 2016a, 2016b, 2017; Hall *et al.*, 2016; Amore *et al.*, 2017; Amore, this volume).

Disasters are human, environmental and economic tragedies (Rose, 2011). The United Nations Office for Disaster Risk Reduction (UNISDR) (2015: 25) defines 'disaster recovery in terms of livelihoods, health, economic, physical, social, cultural and environmental assets, systems and activities'. Earthquakes impact communities in different ways. In terms of the physical environment, earthquakes can cause changes in landform, vegetation and soils, and alterations of hydrological conditions (Migon & Pijet-Migon, 2019). Although subsequent natural processes themselves act towards erasing traces of natural disasters, human interventions can also speed up this process. The physical impacts depend on the hazard mitigation and emergency preparedness practices of the community (Russell *et al.*, 1995; Geschwind, 2001). Both of these, can reduce the physical impacts (Lindell & Prater, 2003). However, they can also induce a number of social impacts that can last for years and decades. The greatest physical impacts relate often to the number of casualties and extent of damage to property and lifeline infrastructures. The extent of the physical impacts is often difficult to assess as casualties may be an indirect consequence of the mainshock or aftershocks. Losses of structures, animals and crops are also important measures of physical impacts (Lindell & Prater, 2003). As argued by Whitman et al. (2013), earthquakes have different impacts on rural and urban areas and studies tend to suggest that the latter recovers faster (Frazier et al., 2013; Cui et al., 2018). Earthquake damage to the built environment can be classified broadly as affecting residential, commercial, industrial, infrastructure or community services sectors (Lindell & Prater, 2003). One way to reduce the physical impacts is to adopt hazard mitigation practices such as avoiding or changing construction in areas that are susceptible to hazard impact. Building construction practices can also make structures less vulnerable (Palm, 1998; Godshalk 2003; Lindell & Prater, 2003; Sengezer & Koç, 2005).

Earthquakes can severely impact organizations in the form of direct physical damage to structures and property, inventory, non-structural damage to premises, changes in cash flow, halted or slowed production, changes in suppliers and customers, staff attrition and psychosocial effects on staff and family (Corey & Deitch, 2011; Whitman et al., 2013). Following the 2010/2011 Canterbury earthquakes, Brown et al. (2015) found that 'customer issues' impacts were the most disruptive for organizations. The disruption of critical services and organizational size work hand in hand with sector-specific organizational vulnerabilities to maximize negative impacts of disasters on organizations (Whitman et al., 2013). Thus, analysing the effects of earthquakes on organizations from a spatial, organizational characteristics and sectoral perspective is a necessary step in improving mitigation strategies that can better inform policy decisions, but also improve organization and community resilience (Whitman et al., 2014).

Earthquakes, Social Impacts and Well-being

The social impacts of disasters can take various forms and includes socio-demographic, socioeconomic, sociopolitical and psychosocial impacts (Lindell & Prater, 2003; Amini Hosseini et al., 2013; Potter et al., 2015; Van der Voort & Vanclay, 2015). One of the most significant socio-demographic impacts of an earthquake-related disaster on a community is the destruction of household dwellings (Mileti & Passerini, 1996). This causes direct economic losses that can be thought of as a loss in asset value but the emotional impacts of losing one's dwelling can be even harder on individuals (Wu & Lindell, 2004; Yi & Yang, 2014; Tierney & Oliver-Smith, 2012). There is also evidence that disaster impacts can cause social activism resulting in political disruption, especially when disaster recovery seems to take longer than what the community anticipates and/or when some interests and groups are excluding from decision-making (Lindell & Prater, 2003; Hall et al.,

2016; Amore *et al.*, 2017), while earthquake recovery and rebuilding process can also be an opportunity for some interests to implement new political structures and advanced particular ideological agendas (Amore & Hall, 2016a, 2017).

Those affected by earthquake disasters often experience a significant decrease in quality of life. Psychosocial impacts are often manifested by psychophysiological effects such as fatigue and tics but also cognitive signs such as confusion, impaired concentration and attention deficits (Tierney & Oliver-Smith, 2012). Emotional signs such as anxiety, depression and grief, as well as behavioural effects such as sleep and appetite changes, ritualistic behaviour and substance abuse are also common (Lindell & Prater, 2003). The earthquake and subsequent nuclear disaster at Fukushima in Japan had not only physical and socioeconomic impacts such as income and job losses, but also psychological and physiological impacts in terms of health impairments such as fear and anxiety related to mental distress from fatalities, injuries or radioactive contamination (Rehdanz *et al.*, 2015). There are also psychosocial impacts with long-term adaptive consequences such as changes in risk perceptions and increased hazard intrusiveness (Lindell & Prater, 2003). Often stories around the recovery of communities following an earthquake emphasize how they overcame physical impacts, paying less attention to psychosocial recovery. In fact, the disaster management literature often portrays well-being of communities as something secondary to the management of physical impacts of earthquakes (Tierney & Oliver-Smith, 2012). However, from a destination and place perspective, both are equally important and deeply intertwined.

The psychological impacts, including well-being, are often understated but require greater attention in disaster management models. Yet, there is no consistency in the literature neither on the extent to which community well-being can be affected nor on the length of time it takes for communities to recover (Prayag *et al.*, 2019a). For example, a study on the psychological adaptation of those affected by the Great Hanshin-Awaji Earthquake of 1995 in Japan, shows that 16 years later, residents with at least one immediate family member who died in the earthquake reported lower life satisfaction, more negative effects and more health problems (Oishi *et al.*, 2015). While it is well established that victims tend to suffer from post-traumatic stress disorder (PTSD) and depressive symptoms, there are also stories of personal growth over time (Lowe *et al.*, 2013). Strong evidence exists to suggest that life satisfaction and well-being decreases substantially in places closer to the disaster (Rehdanz *et al.*, 2015). However, culture has a significant role to play in both recovery of individuals and communities (Palm, 1998) as well as future risk perceptions. Rehdanz *et al.* (2015) found that residents' evaluation of their overall quality of life after the Fukushima disaster was marginally lower and they attributed this to the Buddhism and Daoism philosophical traditions, emphasizing the dialectical nature of things. East Asians, for example, display a high degree of equanimity in the face of negative emotions and events (Rehdanz et al., 2015). However, in a contrary study on the well-being of elderly survivors from the same disaster, Sugano (2016) suggested that psychological well-being and health of survivors changed little compared to pre-disaster levels arguing that the state of the Japanese economy is potentially a reason to explain why life satisfaction was not significantly affected by the earthquake.

At the destination level psychosocial issues during the Canterbury earthquakes in New Zealand have been well documented (Becker et al., 2019). In a review of 31 papers on the psychological impacts of the Canterbury earthquakes on mental health, Beaglehole et al. (2019) found that the mental health of people was affected and strategies needed to be implemented to enable communities to respond to psychological distress. The Canterbury well-being survey which has been ongoing since the initial earthquake, specifically indicated that aftershocks were a major source of anxiety in the greater Christchurch population, with the worst anxiety levels occurring approximately 18 months after the initial mainshock (Morgan et al., 2015). In helping to facilitate community recovery, strategies included community access to free counselling, extended general practice consultations and health promoting initiatives (Beaglehole et al., 2019). The authors argue that these facilities and initiatives had the possible effect of lowering the adverse consequences of the earthquakes on mental health. In addition, social relationships have been found to be the strongest predictor of subjective well-being following disasters (Diener & Seligman, 2002). This issue is discussed in more depth below in relation to earthquakes and resilience.

The existence of prolonged aftershocks that result in communities going through periods of impact, response and recovery several times has not received significant attention in the emergency and disaster management literature (Becker et al., 2019), nor in terms of the effects on tourism (Mazzocchi & Montini, 2001; Mazzoni et al., 2018). As a result, many disaster management models, including those in tourism, tend to be relatively static. Another significant omission in these models is that issues of self-efficacy, empowerment, optimism, innovative thinking, self-esteem, agency, decision making and perceptions are often considered in isolation from business and physical impacts in terms of understanding how people cope with shock, disturbances and stressors (Brown & Westaway, 2011). Increasingly, emphasis needs to be placed in these models on the adaptation of individuals, organizations and communities to changed circumstances, and therefore the idea of bouncing back to a pre-earthquake reality, which is contested in the resilience literature as will be discussed later, should be viewed through a resilience thinking lens (Hall et al., 2016; Hall et al., 2018). These models often also ignore that some individuals and communities can have a 'fresh start mindset', which is the belief that people can make a new start, get a new beginning, and chart a new course in life, regardless of past or present circumstances (Price *et al.*, 2018). This mindset is related to the ability of individuals to choose to reinvent themselves by initiating new goals and adopting new lifestyles to create different futures.

To this end, disaster management and recovery models, and the role of tourism within them, should explicitly account for psychosocial recovery and building resilience. Psychosocial recovery has linkages to the psychological resilience of community members and emotional attachment to places that contribute to psychosocial recovery and which is also very significant in the tourism literature as well, both in terms of residents of a destination as well as visitors (Amsden *et al.*, 2010; Kamani-Fard *et al.*, 2012; Wang *et al.*, 2019). The positive psychology literature abounds with studies arguing that psychological resilience allows individuals to cope with adversity and positively adapt to a changed reality (Kimhi, 2016; Hall *et al.*, 2018). A high level of community resilience enhances individual's coping during stressful situations and is therefore instrumental in faster post-stress recovery (Sherrieb *et al.*, 2010; Chowdhury *et al.*, 2019). Indeed, as Hall *et al.* (2018: 155) concluded

> A resilient community, organisation or destination requires strong interconnectivity. This is similar to individual resilience which is much dependent on formal and informal relationships (Biggs *et al.*, 2012). Nevertheless, social capital requires skilful investment and management for accumulation for use in times of need (Reich, 2006). Therefore, the development of trust between actors and actor engagement and learning are important for resilience (Adger, 2000), especially because when actors trust one another there is an increased likelihood of working towards common goals and outside everyday silos (Hall, 2008). From a tourism perspective, this should not be regarded as a surprising observation; rather it should be standard tourism planning and business practice. Perhaps, as in many things, the focus needs to be not so much on finding new ways to do things but on making sure that the strategies that we know work and helping to ensure that tourism businesses, employees and destinations survive and grow: collaboration; providing a decent standard of living and quality of life for employees and managers; developing trust and talking between actors; and caring about customers, staff and the community.

Earthquakes and Impacts on the Tourism System

Earthquakes have profound effects on all parts of the tourism system. However, the literature examining how tourist destinations, businesses and individuals prepare for, cope, and adjust to disasters is limited (Khazai *et al.*, 2018) and, arguably, that on tourist generating regions, transit regions and competitor destinations and attractions, as well as the tourists themselves, even more so. The focus of earthquake and tourism research tends to be at the destination level. In one sense this is not surprising given that the vulnerability of the tourism industry, at the destination scale, is substantially related to perceptions of safety, functioning infrastructure and visitor accessibility and mobility (Laws & Prideaux, 2005; Hall *et al.*, 2018), all of which can be severely impacted by earthquakes. This perhaps explains the recent effort of many destinations to develop a disaster management plan for the tourism industry, which is vital so that negative impacts can be reduced and recovery time for individuals, communities and destinations improved. Nevertheless, it does not provide a system wide understanding of the effects of a large earthquake related disaster. Furthermore, even at the destination level, often the reaction to a disaster is the development of a disaster management plan rather than proactive decision making by tourism businesses and relevant government departments that incorporate disaster readiness into their daily operations and strategies (Orchiston, 2013; Tsai & Chen, 2010). In many cases, the regions affected by earthquakes, for example, are characterized by high disaster risk but lack sufficient resources for comprehensive public disaster relief work (Tsai & Chen, 2010; see Das & Chakrabarty, this volume).

No destination is immune to natural hazards, thereby requiring destination marketing and management organisations (DMOs) and the tourism industry to work collaboratively with local and central government to develop disaster plans and management strategies (Nguyen *et al.*, 2018). Disasters have an impact on all aspects of the tourism system, including the generating region, transit routes and the destination region (Figure 1.1). However, the full system-wide affects are often not sufficiently appreciated, for example there is very little research of the impact of a disaster at a destination on the transit regions and stops that are connected to it. Nevertheless, as shown in several studies disasters can affect tourism demand (Mazzocchi & Montini, 2001; Huang & Min, 2002; Wang, 2009; Mendoza *et al.*, 2012; Wu & Hayashi, 2014; Cró & Martin, 2017) through tourists' negative perceptions of safety and security, as well as access to accommodation and transport. Often, both air traffic and maritime traffic have to be diverted from the destination if critical infrastructures such as airports and ports have been damaged, requiring travellers to find alternative transit routes (see Morpeth, this volume, for a discussion of the controversies surrounding diversion and tourist access in emergency situations). The impacts of disasters on the destination region is well documented in the literature. It is, therefore, not surprising that several management frameworks focused on disaster response in the context of tourism have been proposed (see Faulkner, 2001; Hystad & Keller, 2008; Ritchie, 2008). Disturbances in one part of the tourism system, e.g. the destination region, also has positive and negative cascading effects on other linked

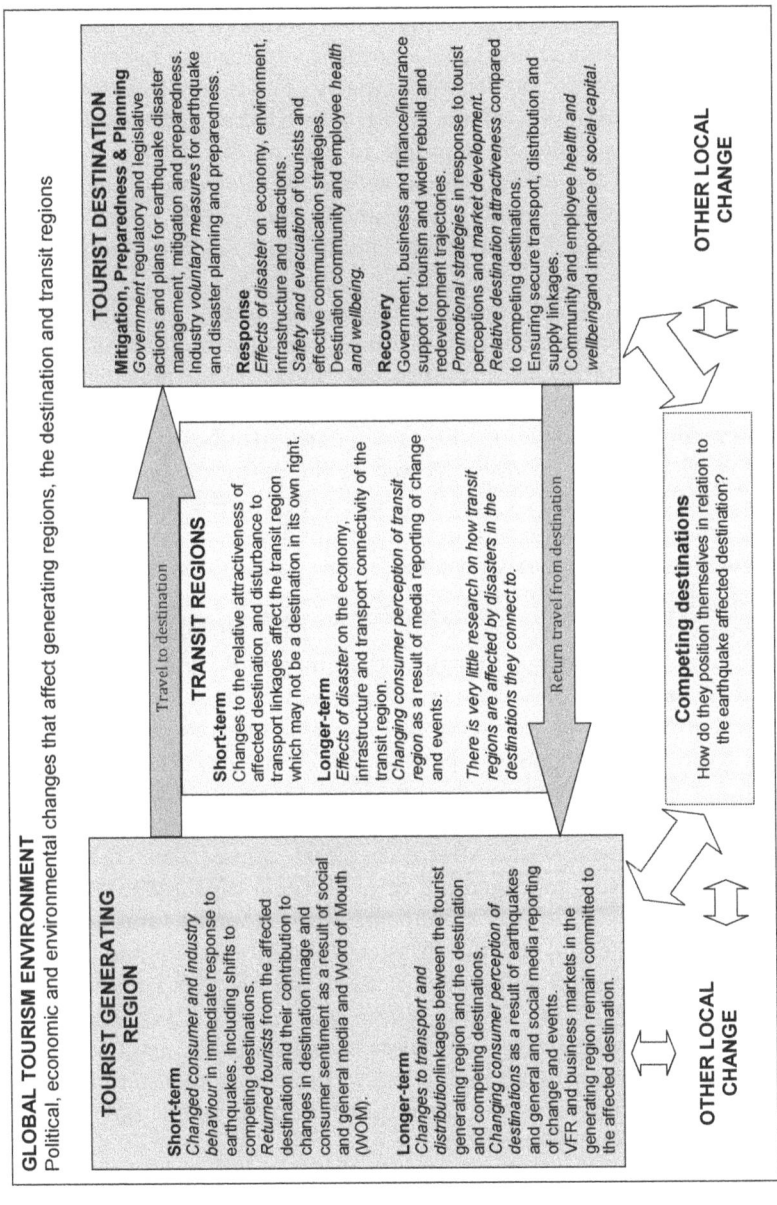

Figure 1.1 System dimensions of tourism in earthquake affected destinations (after Hall, 2005)

parts of the system (Farrell & Twining-Ward, 2004; Hall, 2008). One destinations disadvantage may be advantageous to another as tourists seek to provide substitute holiday experiences. However, several barriers such as adopting disaster preparedness initiatives, including evacuation training, maintaining emergency supplies and communicating hazard risks to tourists have been identified that impede the tourism industry's ability to respond effectively to the negative impacts of disasters (Nguyen *et al.*, 2018; see also Subadra, this volume, and Das & Chakrabarty, this volume). Much of the existing tourism literature focuses on disaster planning and management with greater attention needed to understand the actual recovery process of destinations (Amore & Hall, 2016a, 2016b, 2017; Orchiston & Higham, 2016; Hall & Amore, 2019; see also Amore, this volume).

Blackman *et al.* (2017) argue that there are several effective means for achieving disaster resilience but these often fail on the basis of the purpose of long-term disaster recovery and actual implementation of systems and plans. The most difficult aspects of recovery are to assess the direct impacts of the earthquake, psychosocial wellbeing and perceptions of the recovery as well as the performance of recovery agencies (Bidwell, 2011). Destinations can experience a significant drop in both domestic and international visitor numbers and can lose lifeline infrastructures that stall recovery and reduce accommodation capacity. This can present a challenge to DMOs at both national and regional levels (Orchiston & Higham, 2016). Yet, these challenges are often compounded by governance structures that impede DMOs collaboration with other agencies in the recovery process as well as the overall nature of governance (Amore & Hall, 2016a, 2017). Although rebuilding infrastructure is not necessarily part of their mandate, DMOs can play a significant role in minimizing negative impacts, assisting in defining the roles and responsibilities of tourism stakeholders, and disaster planning and response (Nguyen *et al.*, 2018). Pike (2004) argued that DMOs can contribute to disaster management through their capacity to establish effective media relations, communicate with tourists and visitors, support local businesses, enhance disaster risk awareness among tourism operators and outsource roles when needed (see also Pottorff & Neal, 1994; Drabek, 1999, 2000; Ritchie, 2008; Mair *et al.*, 2016). They may also coordinate specific aspects of disaster management planning from a destination perspective although it is important that such activities are undertaken in conjunction with the responsible government agencies for disaster response (Orchiston, 2013).

Tourism infrastructure does not exist in a vacuum, rebuilding these often require broader considerations of leisure and recreation facilities for residents and consultation with stakeholders that are not necessarily part of the tourism system (Amore & Hall, 2016b). Examples from the Canterbury earthquakes show that two years after the February 2011 earthquake event, residents felt that many factors were still having major negative impacts on their everyday lives. These incudes such things as the inability to make decisions about house damage, repairs and location; being in a damaged environment; loss of recreational, cultural and leisure facilities; additional financial burdens; distress and anxiety associated with aftershocks; loss of usual access to the natural environment and outdoor recreation venues; loss of meeting places for community events; and a lack of opportunities to engage with others in the community through arts, cultural, sport or other leisure pursuits (Morgan et al., 2015; Hall et al., 2016). These issues have significant implications not only for residents quality of life but also affects the tourism industry in terms of, for example, the events venues, art centres and recreation and sport venues that will be rebuilt and for which taxpayers often provide a substantial amount of financial support (Hall & Amore, 2019).

Destination recovery should therefore ideally take more than just an economic perspective that focuses on the restoration of visitor numbers and growth to pre-disaster levels (Hall et al., 2018). There is also a social dimension whereby locals may still experience mental distress even after the tourism destination's economy has recovered (Ritchie, 2009) and which may provide a base for resentment towards tourists and/or the development of tourism infrastructure especially while 'local' infrastructure and needs remain unmet. The local community perceptions and acceptance of tourists, and of tourism as a pathway for recovery, may therefore be considerably different after a disaster than it was before and therefore, community recovery cannot be isolated from the recovery of the tourism industry (Hall et al., 2018). Despite these vulnerabilities, the tourism industry can be reluctant to adopt mitigation strategies of a structural (e.g. investing in reconstruction and maintenance) and non-structural nature (e.g. early warning signs, communication, education and evacuation drills) due to financial reasons (Nguyen et al., 2018) or even a concern with worrying tourists as to the degree of risk. The industry may also not work collaboratively with other sectors and stakeholders to facilitate recovery. In effect, the tourism industry can end up engaging in rather costly, non-financially beneficial, approaches to rebuilding elements of the destination which do not minimize exposure to hazards and actually weakens disaster resiliency over the long term.

Demand-Side Perspectives

The literature on the impacts of disasters on tourists is heavily biased towards estimating tourism demand post-disaster or comparing pre- and post-disaster demand levels. There is a general agreement in this literature that after a disaster, tourism numbers generally decline

(Khazai *et al.*, 2018), with the impacts of earthquakes on tourism numbers examined by several studies (Cró & Martins, 2017; Huang & Min, 2002; Mazzocchi & Montini, 2001). For example, Mazzocchi and Montini (2001) found that the average stay in Umbria, following the central Italy earthquake of September 1997, was increasing due to media reporters and technicians that were covering the disaster staying longer rather than tourists. Examining the impacts of the September 1999 earthquake in Taiwan on the tourism industry, Huang and Min (2002) found that the recovery period exceeded 11 months, with restricted growth in inbound tourist arrivals. In the case of Japan, the impact caused by earthquakes was found to be temporary in nature (Wu & Hayashi, 2014). In the weeks following the Nepal earthquake in April 2015, many tourists were evacuated or departed and forward bookings plummeted (Beirman *et al.*, 2018). Cró and Martins (2017) found no structural breaks in international tourist arrivals for New Zealand. They conclude that this may be due to the earthquake damage being localized in Christchurch. The results of this study were based on national level aggregated data and therefore do not necessarily portray the impacts of the earthquake on the local economy. Prayag *et al.* (2019a) analysed the impact of the 2010 and 2011 Canterbury earthquakes on domestic and international tourism expenditure for Christchurch. They found that the impacts on international tourism was absorbed quicker than that for domestic tourism. They also showed stability in both visitor expenditure and exchange rates post-quake for the international tourism market. This stability can be attributed partly to the response of the tourism industry. As Tucker *et al.* (2017) noted, the marketing and promotion activities of the city, for example, have had an emphasis on rebirth and renewal as opposed to devastation. Altogether, research shows that crises and disasters obviously have some effects on tourism demand but the magnitude of such effects is inconsistent across disasters, locations and time. However, of key importance in understanding the effects of earthquakes on tourist demand is to recognize that media coverage and the framing of earthquake impact at a destination is of much more direct importance to influencing demand than the actual physical impact. Therefore, the development of effective communication strategies by DMOs is of central importance to effectively managing visitor demand following an earthquake (Hall, 2014; Orchiston & Higham, 2016)

Some studies have examined the growth of so-called 'dark tourism' as a segment following a disaster (see Wright, this volume). Studying visitors' perceptions, attitudes and behaviours at earthquake related sites has received some attention in the literature. For example, Yan *et al.* (2016) examine the motivation and emotional reactions of visitors to Sichuan, China, following the Weichuan earthquake. They found that curiosity, leisure related motivations and learning were strong drivers to visit the Weichuan earthquake relics. Hall (2012) criticized the use of the notion of dark tourism in the context of the Christchurch earthquakes as he found that the majority of the domestic tourists he interviewed after the 2010 and 2011 earthquakes were motivated by wanting to better understand the effect of the earthquakes on heritage and on the city as a whole, with the desire to visit substantially influenced by their sense of place and their connection as VFR travellers. An understanding of post-disaster tourist behaviour is increasingly important for future disaster response and planning as tourists are often a key stakeholder affected. Beyond understanding new segments such as dark tourism, an understanding of how specific segments react both psychologically and behaviourally would allow the development and design of strategic marketing communications to mitigate apprehension caused by misperceptions and confusion surrounding the extent of the disaster and its associated risks. This would also facilitate the development and design of strategic marketing communications (Wu & Walters, 2016).

Another emerging segment following a disaster is that of volunteer tourism. For example, in the case of the 2011 Tōhoku earthquake in Japan, Fukui and Ohe (2019) reported a massive influx of volunteer tourists from other prefectures in Japan and internationally but this had fallen substantially six months after the disaster. There is increasing recognition that volunteer tourists can make meaningful contributions to the recovery of communities provided that their activities are appropriately managed. For example, volunteer tourists played a central role in Nepal's tourism recovery following the April 2015 earthquake (Beirman *et al.*, 2018; Wearing *et al.*, 2020). However, the timing of such arrivals on the context of recovery is a significant issue. Another segment that has been highlighted of interest is that of accessible tourism following a disaster. The travel mobility of earthquake survivors as well as making a destination accessible to visitors with disability should be considered as part of the recovery. However, there are limited studies examining the needs of such segments. The study by Tao *et al.* (2019) on the Sichuan earthquake found that survivors would limit their involvement in travel opportunities to avoid public scrutiny. Negative experiences related to perceived unfriendly attitudes of facility staff and displeasure at being stared at by others are common barriers that impact earthquake survivors travel.

The experience of the 2004 earthquake and Indian Ocean tsunami highlights the gap in hazard risk perception between tourists and the local community. Interviews with British tourists visiting locations affected by the disaster revealed that many of them did not evacuate after the initial tremors (Kelman *et al.*, 2008). More disturbing is the fact that nearly half of the fatalities in Khao Lak, Thailand consisted of tourists who had low awareness of local hazard risks and evacuation strategies. This situation highlights the role and responsibilities of tourism and government stakeholders in ensuring tourist safety and security as well as that of local people (Nguyen *et al.*, 2018; see also Subadra, this volume). Many foreign victims of the Asian tsunami had come from low-seismicity countries, where hardly any earthquake induced damage had been documented throughout their written history (Thoresen *et al.*, 2009). Tourist studies should, therefore, focus on understanding the general orientation of tourists toward disaster related risks, awareness of disaster risks and their perceptions of the effectiveness of different emergency response mechanisms, as well as the willingness of local tourism stakeholders to communicate risk and the manner in which hazard risks and evacuation strategies are best conveyed to visitors. Among such issues is the need to present disaster information for tourists in multiple languages (Nguyen *et al.*, 2018).

Supply-Side Perspectives

Much of the research on the impact of disasters in the tourism industry is supply-side oriented (Ritchie, 2008; Wu & Walters, 2016). Many tourism operators are small and micro-enterprises that often lack the resources and capabilities to rebound quickly following a disaster (Mair *et al.*, 2016). They rely on collaborative approaches between national and local government, DMOs and other stakeholders to manage recovery marketing. A key part of managing recovery of tourism destinations is restoring the image and reputation of the place (Khazai *et al.*, 2018). Marketing and promotions are key to assisting a tourist destination to recover after a disaster (Hystad & Keller, 2008). Post-disaster marketing should be aimed at correcting misperceptions and providing information about the recovery phase as well as balancing demand with the capacity to host tourists during a rebuild (Hall, 2014). It is also an opportunity for the destination to correct negative media coverage, if any, about the scale of the disaster, the extent of the damage or the size of the area affected (Mair *et al.*, 2016), as well as the nature of community responses (Carter & Kenney, 2018). The issue of inaccurate media coverage and its effects on destinations and businesses following a crisis has been highlighted in previous studies (Sonmez *et al.*, 1999). For example, initial international reporting of the Kaikoura earthquake in 2016 suggested that the quake had hit north-east of Christchurch, which immediately resonated with the public as another earthquake in Christchurch, although it was Kaikoura that was substantially affected as a destination (Fountain & Cradock-Henry, 2019; see also Fang *et al.*, this volume). The further the generating market is from the affected destination, the more vulnerable it seems to be to sensationalized and inaccurate media coverage, which is often the root cause of negative perceptions about a destination (Hall, 2010; Walters & Clulow, 2010).

The timing of recovery marketing efforts should be a collective one, across local, regional and national tourism stakeholders. More

importantly questions about the appropriateness, ethics, timing and effectiveness of different recovery phase marketing strategies need to be addressed (Orchiston & Higham, 2016). As an example, Tourism New Zealand (TNZ), the central government agency for marketing New Zealand internationally as a tourist destination, removed images of Christchurch from all international marketing material after the 2011 February earthquake. In parallel, existing advertising from news websites were removed and key word searches associated with the Canterbury earthquakes were purchased to deflect web browsers from negative imagery and to promote positive searches for New Zealand. For many countries, media monitoring is an essential part of disaster management planning for tourism, allowing destinations to counteract any negative publicity, thereby limiting damage to destination image and reputation (Huang *et al.*, 2008). In addition, all internet traffic to the nz.com website was directed to the corporate website so that all communication related to the earthquake could be separated from tourism promotion of New Zealand. Both TNZ and Christchurch and Canterbury Tourism (CCT) agreed that a period of demarketing was necessary (Orchiston & Higham, 2016). From September 2011 onwards, several campaigns such as the 'South Island Road Trips' and 'Christchurch Reimagined' followed as part of recovery marketing efforts (Orchiston & Higham, 2016). Disseminating positive new stories can also be very effective at offsetting negative destination publicity caused by mass media reporting (Chacko & Marcell, 2008), while select use of social media has also become increasingly important with respect to post-disaster destination image management.

The lack of research with respect to the communication strategies used by stakeholders in times of crisis and disasters has become increasingly recognized (Mair *et al.*, 2016; Seyfi & Hall, 2020). Studies tend to examine the role of post-disaster recovery marketing messages in the form of information provision to tourists and recovery slogans. 'Open for business' is, for example, a common theme for post-disaster recovery marketing messages (Prideaux *et al.*, 2008). 'Nepal Back on the Top of the World' was used as a recovery slogan to symbolize repositioning of Nepal from victimhood to restoration (Beirman *et al.*, 2018). The DMO also employed celebrity visits to give prominence and visibility to the campaign. Accessible tourism enterprises were provided opportunities for them to create their own narratives and to include them in the broader media and marketing approach towards stimulating tourism recovery (Beirman *et al.*, 2018). The role of Twitter and other social media in promoting resilience (Veer *et al.*, 2016) and destination recovery has also been given increased attention (Fukui & Ohe, 2019), with evidence suggesting that social media can help with recovery but that it can also potentially fuel negative perceptions about the destination.

Risk communication is not only aimed at tourists but should also take into account the communication needs of residents. Strategies should be adapted to fit into the disaster recovery process to accommodate the changing and evolving challenges of residents (e.g. relocation and progressive damage) (Deng et al., 2017; Subedi et al., 2018; Becker et al., 2019). For example, anxiety caused by the aftershocks following the February 2011 Canterbury earthquakes had implications for people's interpretation and sense-making of earthquake information. The ability of people to understand what was happening in the city diminished because of stress and anxiety (Veer et al., 2016; Becker et al., 2019), and therefore, the effectiveness of messaging from disaster management authorities began to be questioned. Becker et al. (2019) provide several recommendations about post-earthquake risk communication including the need to have a clear communication strategy prior to an earthquake, allowing for flexibility in communication and providing training and education about aftershocks to both tourists and residents. There is a therefore clear need for better matching of information needs to different audiences/market segments following a disaster, with Becker et al. (2019) highlighting the need to inject empathy in aftershock communication and to ensure inter-agency coordination around communication, among others.

While understanding the behaviour of new segments such as dark tourism is important, there are also issues surrounding whether such segments should be actively promoted (see Wright, this volume). Disaster tourism is a distinctive form of dark tourism in that the local community often becomes the focus of the disaster tourist gaze (Wright & Sharpley, 2018). Often disaster tourism sites develop without the destination deliberately embarking on the development of such sites. Unlike the (re)construction of commemorative sites that can become a new tourism resource that can generate revenue or improve the attractiveness of the destination, for example interpretation of post-earthquake heritage conservation, dark tourism is a double-edged sword for destinations. On the one hand, it can attract a significant number of tourists. On the other, it can become a source of conflict between the tourism industry and residents. The negative narratives of loss associated with dark tourism can cause residents to reject support for the development of such sites but may also prevent psychosocial recovery. It has been argued that the narratives around such sites must be transformed into positive accounts of communal renewal and hope. For example, Lin et al. (2018) coined the term 'blue tourism' as a community led approach to post-disaster tourism development. Blue tourism is described as a form of resilience which builds around local place-based practices and traditional community knowledge. This approach, rather than dark tourism, is supposedly capable of achieving sustainable disaster recovery and tourist satisfaction simultaneously and potentially offers a more nuanced understanding of the role of community-based tourism initiatives in enhancing resilience and pursuing a more sustainable form of tourism in post-disaster areas.

Post-disaster recovery very often is focused on removing the tangible evidence and rebuilding of damaged objects, which is understandable and expected from the perspective of affected communities (Migon & Pijet-Migon 2019). The end result can be the gradual disappearance of the event from human memory. Examining selected Italian disasters, Coratza and De Waele (2012) underlined the importance of such sites for earthquake and geological education. Leaving some of the evidence of disasters may have positive effects for learning, understanding, and adding to the recovery of affected communities and can also serve the geo-tourism segment (Migon & Pijet-Migon, 2019). However, Migon and Pijet-Migon (2019) argue that sites focused on the disrupted lives of communities while honouring victims contribute very little to improving understanding of vulnerability and risk. Instead, they propose that the development of thematic trails around the disaster location showcasing various aspects of the disaster linked with source/cause/effect and the topographic context can potentially be a better way to keep the event from disappearing from the collective human memory.

Earthquakes and Resilience

The growth in the human impact of disasters has strengthened a focus in research and policy in understanding preparedness and response to hazard events (Thompson *et al.*, 2017). These are often framed in terms of such notions as a disaster response or planning cycle with each stage informing the next such as that of preparedness, response, recovery, planning post-event (Gurwitch *et al.*, 2004) or rebuilding, redeveloping and renewal to support effective recovery (Blakely, 2012) or the adoption of the four stage approach of mitigation, preparedness, response and recovery (Hernantes *et al.*, 2013). There is often an implicit assumption in these frameworks that once the community goes back to their normal lives or plans have been put in place to mitigate the effects of future similar disasters, then the process ends (Muskat *et al.*, 2015). It is also implied that a community or destination will move from one part of the cycle to the next in a linear fashion and is almost automatic (Muskat *et al.*, 2015). Yet, as evidenced by disasters such as Hurricane Katrina, the Christchurch and Nepal earthquake and nuclear meltdown in Fukushima, systems, cycles, and plans can fail and there is a need to build adaptive capacity and resilience (Hall *et al.*, 2018).

Resilience has been conceptualized as the capacity for communities and their members, including businesses and societal institutions to respond to crises and disasters (Paton, 2008). The resilience literature is grounded in an understanding of the dynamics of change, complexity, the potential role of transitions and the possibility of crises providing windows of opportunity (Brown & Westaway, 2011; Hall et al., 2018; Amore et al., 2018). These are some of the underlying premises of social-ecological systems. Adaptive capacity, resilience and vulnerability are related and entwined in different ways (Engle, 2011). Resilience is commonly explained through an adaptive cycle. This cycle does not converge to a state of equilibrium but rather moves through states of growth, conservation, collapse and re-organization (Holling & Gunderson, 2002). This enables a system to harness transformative or adaptive capabilities to address change and maintain a cyclical process (Bec et al., 2016).

Resilience frameworks in tourism have generally been adapted from those in other disciplines (Cochrane, 2010). Irrespective of the frameworks, it is clear that pro-active tourism policymaking, planning and implementation of disaster risk reduction are likely to enhance the sector's ability to recover from crises and disasters (Khazai et al., 2018). Understanding the vulnerability of a destination is the starting point for resilience building activities. Examples of destination vulnerability include limited disaster preparedness, access to resources, being ecologically sensitive and hazard prone and suffering from institutional inflexibility, among others (Calgaro et al., 2014). Amore et al. (2018) use a multilevel perspective to argue that destination planning frameworks, and hence destination resilience building, should encompass ecological, socioecological, sociopolitical, socioeconomic and sociotechnological dimensions that reflect the embeddedness of resilience among heterogeneous and potentially complementary destination stakeholders. They highlight that a resilience approach to destination planning offers destinations not only the possibility of coping with sudden changes such as disasters but also incremental changes, which is part of a business as usual approach. For example, participatory approaches in crafting a disaster management plan where stakeholders beyond the tourism industry understand and are willing to share resources, knowledge and information, can lead to quicker response following a disaster. Also, efforts to build organizational and community resilience can contribute to destination resilience and vice versa (Hall et al., 2018). As Cutter et al. (2013) suggest, disaster resilience is very much linked to collaborative engagement across organizations. Recovery, in particular, requires multi-agency partnerships and collaboration. Therefore, participatory approaches improve the chance that a disaster management plan has stakeholder buy in, which improves the likelihood of the plan working following a disaster. Several studies have highlighted how organizations can become limited within their silos or lack networked communication practices for sharing best practice (Seville, 2018), which impede not only emergency services as first responders after a disaster but also the tourism industry to initiate, for example, the evacuation of tourists (see Subadra, this volume).

In a resilient socioecological system, disturbance has the potential to create new opportunity for innovation and development (Folke, 2006). Appreciating the dynamic and cross-scale interplay between abrupt change and sources of resilience makes it apparent that the resilience of complex adaptive systems is not simply about resistance to change and conservation of existing structures. It is also about the opportunities that disturbance opens up in terms of the recombination of evolved structures and processes, system renewal and the emergence of new trajectories (Folke, 2006). It is not about returning to normality but about positively adapting to a changed reality. For example, the local economy in Kaikoura, following the 2016 earthquake was revitalized and regional resilience enhanced through diversification, capitalizing on the region's natural, social and cultural capital (Cradock-Henry *et al.*, 2018). In the case of Kaikoura, food security emerged as an important concern for the community post-quake. This led to greater levels of self-organization, in which individuals, households, businesses and rural and urban communities, harnessed local opportunities and connectivity to become food self-reliant (Cradock-Henry *et al.*, 2018). This is an example of building community resilience by capitalizing on the new opportunities presented by the disaster and this is why the concept of resilience incorporates the ideas of adaptation, learning and self-organization in addition to the general ability to persist post-disturbance (Folke, 2006).

Calgaro *et al.* (2014) claim that a lack of understanding of the factors that build and affect destination resilience and vulnerability lead to an inability to effectively build community resilience. Similarly, Pizzo (2015) warns that the notion of resilience is becoming a buzzword and argues that after an unexpected event, not all communities have to be resilient nor should they be resilient to every unexpected event, nor should they be resilient in the same way as a previous similar event. Therefore, communities are not always looking for a new equilibrium, nor are they looking simply to bounce back to their pre-disaster state, especially if the state was less than desirable to begin with (Cowell, 2013). Community – and, hence, destination – resilience is, thus, not the sum of individuals and organizations being resilient (Cutter *et al.*, 2014; Pizzo, 2015). Although these can help to build community and destination resilience (Prayag, 2018), the role of place or neighbourhood in developing social networks for a community's disaster preparedness, response and resilience appears critical (Cox & Perry, 2011; Biggs *et al.*, 2012, 2015; Aldrich & Meyer, 2015; Masterson *et al.*, 2017). The sense of connectedness to the new changed reality is important not only for individuals but also for the community as a whole. Hence, the importance of place attachment for disaster recovery has become increasingly emphasized (Guo *et al.*, 2018). As an example, owner-operators of lifestyle tourism enterprises can develop emotional attachment to their businesses and the associated sense of place, making them more reluctant to abandon the business and the location in difficult times, thus strengthening their resilience in the face of disasters (Biggs *et al.*, 2015).

Related to the above, the literature clearly pinpoints to social capital as an enabler of organizational and community resilience (Biggs *et al.*, 2012, 2015; Hall *et al.*, 2018; Chowdhury *et al.*, 2019). Aldrich (2012) after an extensive review of disaster recovery related to, for example, the Kobe 1995 earthquakes, the 2004 Indian Ocean tsunami and 2005 Hurricane Katrina argued that social capital serves as a core engine of disaster recovery. Social capital is the goodwill engendered by the fabric of social relations that can be mobilized to facilitate action (Adler & Kwon, 2002). Different forms of social capital such as bonding, bridging and linking emerge in the different disaster relief, rehabilitation and recovery phases and play different roles towards overall long term recovery (Blackman *et al.*, 2017). Kinship networks encourage cohesion, connectedness, reassurance and stability in times of need. They also facilitate access to financial capital and power networks (Calgaro *et al.*, 2014). Social sources of resilience such as social capital, which is grounded in trust and social networks and social memory (the experience of dealing with change) are essential for the capacity of socioecological systems to adapt and shape change (Folke *et al.*, 2005). Disaster relief work should, therefore, provide instrumental, informational and emotional support to community members through facilitating them to seek out others and establish bonds with people they know and even strangers. This is the fundamental premise of social relationships post-disaster (Reich, 2006), and, hence the building of social capital. A resilient community or destination is an inter-connected community (Allenby & Fink, 2005). Opportunities to build capacity and capability though the acquisition of new skills, and knowledge sharing would therefore enhance community resilience (Cradock-Henry *et al.*, 2018).

A resilient tourism organization adjusts its operations, management and marketing strategies to sustain under dramatically changing conditions (Dahles & Susilowati, 2015). In the case of the Canterbury earthquakes, Chowdhury *et al.* (2019) showed that different forms of social capital such as structural, relational and cognitive capital are important but only relational capital had a significant influence on adaptive resilience of tourism organizations. The importance of social capital can also be seen in its direct impact on the financial performance of tourism organizations (Prayag *et al.*, 2018). Both social capital and resilience require trust from actors. If actors trust each other they are more likely to collaborate beyond the restrictions of hierarchical organizations and daily routines (Rogers *et al.*, 2016). The 'silo effect' of inter and intra-organizations often negatively impacts their ability to effectively respond to disasters. The ability to find alternative resources is critical to the resilience of tourism organizations (Dahles & Susilowati, 2015) and communities. Organizational resilience has been described as the inherent characteristics of organizations that are able to respond more quickly, recover faster or develop more unusual ways of doing business under duress compared to others (Sutcliffe & Vogus, 2003). Therefore, tourism organizations need resources to bounce back but often these reside outside the boundaries of the community and require actors' trust of each other and the political processes to share such resources. For business recovery following the Canterbury earthquakes, several national government led initiatives were put in place to support all businesses, including tourism. For example, the earthquake subsidy scheme was introduced by the government, available immediately after the February 2011 earthquake, vouchers were provided to help build capability in earthquake affected businesses, business mentors were appointed to help them to bounce back, and Red Cross grants were available to help businesses to access legal, financial and engineering advice (EQ Recovery Learning, 2016).

Resilience takes place at multiple levels, individual, organizational and community, but few studies explore how the different levels interact (Sutcliffe & Vogus, 2003; Linnenluecke, 2017). In the context of the Canterbury earthquakes, Prayag *et al.* (2019b) showed that in small and medium tourism enterprises, the psychological resilience of owners and managers have an influence on employee resilience, which in turn positively impacts organizational resilience. They highlight that life satisfaction has a role to play post-disaster in building organizational resilience. This aligns with previous studies suggesting that a lack of resilience at one level can undermine resilience at other levels (Hall *et al.*, 2018; Pizzo, 2015). Although research and policy highlight the need to understand human factors in determining adaptive capacity, these are seldom integrated in current disaster models and frameworks (Amore *et al.*, 2018). Individual, family and community characteristics that build resilience, as part of the so-called healthy functioning adaptive systems that support them, are often reduced to minor factors affecting the recovery strategies of communities (Brown & Westaway, 2011). Nevertheless, a significant body of research exists around the resilience of individuals and the role of psychological capital in facilitating recovery from disasters but these have not been integrated adequately in current disaster management models. After disasters people need to believe that they have the personal resources to achieve goals such as rebuilding their homes and businesses, getting jobs and starting their lives again. The psychological key to rebounding is the effort to regain personal control. Disaster planning for tourism, should therefore provide pathways for allowing community members, including business owners or managers and employees to re-establish personal control (Reich, 2006) and facilitate the process of building psychological resilience. Luthans *et al.* (2006), for example, suggest that the resilience of employees can be developed through organizational interventions, for instance, by asking employees to identify personal setbacks within their work domain, to assess the realistic impact of their setback and to identify options for taking action. Resilience in this context is seen as a contributing factor towards employee psychological capital (Linnenluecke, 2017). However, there is an urgent need for studies examining psychological capital in tourism organizations and communities dependent on tourism (Hall et al., 2018; Prayag, 2018).

Resilience and sustainability

Resilience is usually used in the context of coping with change and responding to specific shocks and this relates to short-term survival and recovery (Rose, 2011). Sustainability revolves around long-term survival and improving the quality of life and the environment. In the tourism literature, there is considerable emphasis on resilience to immediate challenges but there is also merit in conceptualizing resilience as a dynamic long-term state, highlighting the obvious parallels with the concept of sustainability (Espiner et al., 2017). Both resilience and sustainability have been described as highly abstract and multifaceted concepts, each with a variety of definitions and interpretations (Derissen et al., 2011; Hall et al., 2018). Resilience is often viewed in normative terms as a need or commitment to become more resilient, similar to the concept of sustainability (Hall et al., 2018). Derissen et al. (2011) proposed four different potential relationships between the two concepts: resilience of a system is necessary but insufficient for sustainability, resilience of a system is sufficient but not necessary for sustainability, resilience of the system is neither necessary nor sufficient for sustainability and resilience of a system is both necessary and sufficient for sustainability. However, the capacity for organizations and destinations to be agile and adaptive in responding to rapid, unexpected change is one clear point of difference between the concepts of resilience and sustainability (Espiner et al., 2017).

Espiner et al. (2017) argue that systems can be resilient without being sustainable, while if a system is sustainable, it is implicitly assumed that it is resilient to change. Destinations cannot be sustainable if they are also not resilient (Espiner et al., 2017). This is particularly the case when destinations are hit by disasters. The initial focus is usually on rebuilding the 'hard' infrastructure such as roads, sewage and airports to enable locals to have access to amenities and social infrastructure. However, consideration must also be given to the long-term needs of the community and its ability to cope with other sudden as well as incremental change. Yet, consideration of social infrastructure related to health and wellbeing is often secondary, while the tourism industry often even fails to provide a living wage in many situations, with both of these factors (among others) having significant implications for both community and destination resilience and sustainability (Pizzo, 2015; Lew *et al.*, 2016; Hall *et al.*, 2018).

Different disasters require different spatial and time frames for policies and action (Pizzo, 2015). Though the two are not mutually exclusive, resilience studies in tourism seem to almost suggest that the same resilience building approach can be applied to every type of disaster. Also, resilience thinking often lacks depth in analysing the social dimension, including the political economy of resource and power distribution, and the consequences of uneven patterns of resource use over space and time (Miller *et al.*, 2010). For example, although social capital is described as a positive resource that allows individuals and communities to cope and bounce back, social relationships and networks can also foster social exclusion in the rebuilding process, manifested through dominant power structures and historically embedded cultural norms. The dark side of social capital must be acknowledged. This was evident across the 2004 tsunami affected destinations in Thailand (Calgaro & Lloyd, 2008). In addition, many adaptive management strategies fail to be successfully implemented or bring about transformative changes due to existing governance structures (Folke *et al.*, 2010). What appears to be a resilient structure can hold power structures, inequities and exclusion in a place that can create rigidity traps (Folke *et al.*, 2010) and lead to substantial questions about the validity of any earthquake disaster responses.

Framework for the Book

As noted in the above discussion, the understanding of disasters can be framed in terms of planning or policy cycles and the response to hazard events (Hall, 2002, 2010; Gurwitch *et al.*, 2004; Blakely, 2012; Hernantes *et al.*, 2013; Thompson *et al.*, 2017). This approach has been used to position the chapters in the present volume in terms of mitigation, preparedness, response and recovery, with the latter stage including aspects of rebuilding, redeveloping and renewal, as well as post-event planning which brings the process full circle but aiming to mitigate future events (Table 1.2). However, it should be noted that many aspects of these stages are not discrete and key concepts, such as social capital, communication and trust in institutions, run across all stages. Chapter 2 by Martini and Platania looks at how resilience and preparedness of tourism operators in Mount Etna is shaped by the experience of previous earthquake events, perception of risk and institutional trust and social capital. Some similar themes are picked up in Chapter 3 by Das and Chakrabarty in their discussion of communication systems and earthquake preparedness for the tourism sector in Nepal. Subadra (Chapter 4) examines the mitigation

26 Tourism and Earthquakes

Table 1.2 Major focus of chapters in relation to stages of disaster management cycle

Chapter	Location	Mitigation	Preparedness	Response	Recovery	Post event planning
2 Martini & Platania	Mount Etna, Italy	Shaped by previous events	Risk perception; social capital			Resilience
3 Das & Chakrabarty	Nepal	Communication to communities and disaster management planning				
4 Subadra	Bali, Indonesia	Communication of risk to tourists and disaster management planning				
5 Amore	Christchurch, New Zealand			Response of tourism-relevant stakeholders		
6 Morpeth	Haiti			Ethics of disaster tourism		
7 Hashimoto & Telfer	Kumamoto, Japan			Industry assistance to evacuees	Industry recovery	
8 Mardiah et al.	Jogjakarta, Indonesia			Role of government and institutional interventions; development of a recovery network		Ongoing issues of risk perceptions
9 Fang et al.	Kaikōura, New Zealand				Individual, organisation and destination resilience	
10 Wright	L'Aquila, Italy				Development of ghost tourism	

of earthquake and tsunami risk in coastal Bali and reinforces the importance of effective communication of risk and appropriate response to tourists.

Chapters 5 to 7 are primarily focused on the response to earthquakes. Chapter 5 by Amore examines the expectations and disillusionment of tourism-relevant stakeholders in Christchurch following the 2010–2011 earthquake sequence and how this affects the trajectory of post-earthquake recovery. Chapter 6 by Morpeth discusses response in a more compressed time frame in discussing the implications of the role of tourism in the immediate aftermath of the 2010 Earthquake in Haiti. In Chapter 7 Hashimoto and Telfer provide a more extended coverage of tourism industry post-disaster response, in their overview of how tourism businesses assisted evacuees after the 2016 Kumamoto earthquakes in Japan. The three chapters taken together are also interesting because of the different perspectives they provide of the role that tourism can play in the aftermath of earthquake related disasters.

Chapters 8 to 10 move from the response stage to more of a focus on recovery. Chapter 8 by Mardiah *et al.* examines the contribution of handicraft shopping tourism to economic recovery after the 2006 Jogjakarta Earthquake, with particular focus on the nature of government and institutional interventions, the development of a recovery network, and ongoing issues of risk perceptions. Chapter 9 by Fang *et al.* looks at the interrelationships between recovery and resilience and their interplay at individual, organizational and destination scales with respect to tourism entrepreneurs in Kaikōura, New Zealand, following the November 2016 earthquakes. Finally, Chapter 10 by Wright looks at the controversial issue of post-earthquake 'dark' and 'ghost' in L'Aquila, Italy. The three chapters collectively highlight the complexity of the tourism system's response to earthquake disasters and the difficult relationships that may develop between tourism and non-tourism recovery goals. The final chapter by Prayag and Hall (Chapter 11) reinforce the major themes of the book and highlight some of the significant research gaps that exist in research on tourism and earthquakes.

Earthquakes and tsunamis undoubtedly have a major impact on communities and destinations and those that experience them and perceptions of risk and place. One of the great weaknesses of much tourism research on the effects of earthquakes is that it only examines a particular moment of the earthquake disaster, response and recovery cycle. Comprehensive long-term overviews are limited or are still being put into place. However, the present assembly of chapters hopefully at least provide some indication of the ongoing response of tourism, tourists and communities to the challenges that earthquakes pose in at-risk destinations and the social, economic and environmental benefits that improved planning and preparedness may bring.

References

Adger, W.N. (2000) Social and ecological resilience: are they related?. *Progress in Human Geography* 24 (3), 347–364.
Adler, P.S. and Kwon, S.W. (2002) Social capital: Prospects for a new concept. *Academy of Management Review* 27 (1), 17–40.
Aldrich, D.P. (2012) *Building Resilience: Social Capital in Post Disaster Recovery*. Chicago: University of Chicago Press.
Aldrich, D.P. and Meyer, M.A. (2015) Social capital and community resilience. *American Behavioral Scientist* 59 (2), 254–269.
Allenby, B. and Fink, J. (2005) Toward inherently secure and resilient societies. *Science* 309 (5737), 1034–1036.
Amini Hosseini, K., Hosseinioon, S. and Pooyan, Z. (2013) An investigation into the socioeconomic aspects of two major earthquakes in Iran. *Disasters* 37 (3), 516–535.
Amore, A. and Hall, C.M. (2016a) From governance to meta-governance in tourism? Re-incorporating politics, interests and values in the analysis of tourism governance. *Tourism Recreation Research* 41 (2), 109–122.
Amore, A. and Hall, C.M. (2016b) 'Regeneration is the focus now': Anchor projects and delivering a new CBD for Christchurch. In C.M. Hall, S. Malinen, R. Vosslamber and R. Wordsworth (eds) *Business and Post-disaster Management: Business, Organisational and Consumer Resilience and the Christchurch Earthquakes* (pp. 181–199). Abingdon: Routledge.
Amore, A. and Hall, C.M. (2017) National and urban public policy in tourism. Towards the emergence of a hyperneoliberal script? *International Journal of Tourism Policy* 7 (1), 4–22.
Amore, A., Hall, C.M. and Jenkins, J. (2017) They never said 'Come here and let's talk about it': Exclusion and non-decision-making in the rebuild of Christchurch, New Zealand. *Local Economy* 32 (7), 617–639.
Amore, A., Prayag, G. and Hall, C.M. (2018) Conceptualizing destination resilience from a multilevel perspective. *Tourism Review International* 22 (3–4), 235–250.
Amsden, B.L., Stedman, R.C. and Kruger, L.E. (2010) The creation and maintenance of sense of place in a tourism-dependent community. *Leisure Sciences* 33 (1), 32–51.
Beaglehole, B., Mulder, R.T., Boden, J.M. and Bell, C.J. (2019) A systematic review of the psychological impacts of the Canterbury earthquakes on mental health. *Australian and New Zealand Journal of Public Health* 43 (3), 274–280.
Bec, A., McLennan, C.L. and Moyle, B.D. (2016) Community resilience to long-term tourism decline and rejuvenation: A literature review and conceptual model. *Current Issues in Tourism* 19 (5), 431–457.
Becker, J.S., Potter, S.H., McBride, S.K., Wein, A., Doyle, E.E.H. and Paton, D. (2019) When the earth doesn't stop shaking: How experiences over time influenced information needs, communication, and interpretation of aftershock information during the Canterbury Earthquake Sequence, New Zealand. *International Journal of Disaster Risk Reduction* 34, 397–411.
Beirman, D., Upadhayaya, P.K., Pradhananga, P. and Darcy, S. (2018) Nepal tourism in the aftermath of the April/May 2015 earthquake and aftershocks: Repercussions, recovery and the rise of new tourism sectors. *Tourism Recreation Research* 43 (4), 544–554.
Bidwell, S. (2011) *Designing Indicators for Measuring Recovery From Disasters*. Christchurch, New Zealand: Canterbury District Health Board.
Biggs, D., Hall, C.M. and Stoeckl, N. (2012) The resilience of formal and informal tourism enterprises to disasters: Reef tourism in Phuket, Thailand. *Journal of Sustainable Tourism* 20 (5), 645–665.
Biggs, D., Hicks, C.C., Cinner, J.E. and Hall, C.M. (2015) Marine tourism in the face of global change: The resilience of enterprises to crises in Thailand and Australia. *Ocean & Coastal Management* 105, 65–74.

Blackman, D., Nakanishi, H. and Benson, A.M. (2017) Disaster resilience as a complex problem: Why linearity is not applicable for long-term recovery. *Technological Forecasting and Social Change* 121, 89–98.

Blakely, E.J. (2012) *My Storm: Managing the Recovery of New Orleans in the Wake of Katrina*. Philadelphia: University of Pennsylvania Press.

Bonati, S. (2015) Multiscalar narratives of a disaster: From media amplification to Western participation in Asian tsunamis. *Culture Unbound: Journal of Current Cultural Research* 7 (3), 496–511.

Brown, C., Stevenson, J., Giovinazzi, S., Seville, E. and Vargo, J. (2015) Factors influencing impacts on and recovery trends of organisations: Evidence from the 2010/2011 Canterbury earthquakes. *International Journal of Disaster Risk Reduction* 14, 56–72.

Brown, K. and Westaway, E. (2011) Agency, capacity, and resilience to environmental change: Lessons from human development, well-being, and disasters. *Annual Review of Environment and Resources* 36, 321–342.

Calgaro, E. and Lloyd, K. (2008) Sun, sea, sand and tsunami: Examining disaster vulnerability in the tourism community of Khao Lak, Thailand. *Singapore Journal of Tropical Geography* 29 (3), 288–306.

Calgaro, E., Lloyd, K. and Dominey-Howes, D. (2014) From vulnerability to transformation: A framework for assessing the vulnerability and resilience of tourism destinations. *Journal of Sustainable Tourism* 22 (3), 341–360.

Carter, L.H. and Kenney, C.M. (2018) A tale of two communities: B-race-ing disaster responses in the media following the Canterbury and Kaikōura earthquakes. *International Journal of Disaster Risk Reduction* 28, 731–738.

Chacko, H.E. and Marcell, M.H. (2008) Repositioning a tourism destination: The case of New Orleans after hurricane Katrina. *Journal of Travel & Tourism Marketing* 23 (2–4), 223–235.

Chowdhury, M., Prayag, G., Orchiston, C. and Spector, S. (2019) Post-disaster social capital, adaptive resilience and business performance of tourism organizations in Christchurch, New Zealand. *Journal of Travel Research* 58 (7), 1209–1226.

Cochrane, J. (2010) The sphere of tourism resilience. *Tourism Recreation Research* 35 (2), 173–185.

Coratza, P. and De Waele, J. (2012) Geomorphosites and natural hazards: Teaching the importance of geomorphology in society. *Geoheritage* 4 (3), 195–203.

Corey, C.M. and Deitch, E.A. (2011) Factors affecting business recovery immediately after Hurricane Katrina. *Journal of Contingencies and Crisis Management* 19 (3), 169–181.

Cowell, M.M. (2013) Bounce back or move on: Regional resilience and economic development planning. *Cities* 30, 212–222.

Cox, R.S. and Perry, K.M.E. (2011) Like a fish out of water: Reconsidering disaster recovery and the role of place and social capital in community disaster resilience. *American Journal of Community Psychology* 48 (3–4), 395–411.

Cradock-Henry, N., Fountain, J. and Buelow, F. (2018) Transformations for resilient rural futures: The case of Kaikōura, Aotearoa-New Zealand. *Sustainability* 10 (6), 1952.

Cró, S. and Martins, A.M. (2017) Structural breaks in international tourism demand: Are they caused by crises or disasters? *Tourism Management* 63, 3–9.

Cui, K., Han, Z. and Wang, D. (2018) Resilience of an earthquake-stricken rural community in southwest China: Correlation with disaster risk reduction efforts. *International Journal of Environmental Research and Public Health* 15 (3), 407.

Cutter, S.L., Ash, K.D. and Emrich, C.T. (2014) The geographies of community disaster resilience. *Global Environmental Change* 29, 65–77.

Cutter, S.L., Ahearn, J.A., Amadei, B., Crawford, P., Eide, E.A., Galloway, G.E. and Scrimshaw, S.C. (2013) Disaster resilience: A national imperative. *Environment: Science and Policy for Sustainable Development* 55 (2), 25–29.

Dahles, H. and Susilowati, T.P. (2015) Business resilience in times of growth and crisis. *Annals of Tourism Research* 51, 34–50.

Daniell, J.E., Khazai, B., Wenzel, F. and Vervaeck, A. (2011) The CATDAT damaging earthquakes database. *Natural Hazards and Earth System Sciences* 11, 2235–2251.

Deng, Y., Su, G., Gao, N. and Sun, L. (2017) Investigation and analysis of the importance awareness of the factors affecting the earthquake emergency and rescue in different areas: A case study of Yunnan and Jiangsu Provinces. *International Journal of Disaster Risk Reduction* 25, 163–172.

Derissen, S., Quaas, M.F. and Baumgärtner, S. (2011) The relationship between resilience and sustainability of ecological-economic systems. *Ecological Economics* 70 (6), 1121–1128.

Diener, E. and Seligman, M.E. (2002) Very happy people. *Psychological Science* 13 (1), 81–84.

Drabek, T.E. (1999) Disaster evacuation responses by tourists and other types of transients. *International Journal of Public Administration* 22 (5), 655–677.

Drabek, T.E. (2000) Disaster evacuations: tourist-business managers rarely act as customers expect. *Cornell Hotel and Restaurant Administration Quarterly* 41 (4), 48–57.

Engle, N.L. (2011) Adaptive capacity and its assessment. *Global Environmental Change* 21 (2), 647–656.

EQ Recovery Learning (2016) *Earthquake Recovery in Canterbury* [online]. See https://www.eqrecoverylearning.org/about/earthquake-recovery-in-canterbury/ (accessed June 2019).

Espiner, S., Orchiston, C. and Higham, J. (2017) Resilience and sustainability: A complementary relationship? Towards a practical conceptual model for the sustainability–resilience nexus in tourism. *Journal of Sustainable Tourism* 25 (10), 1385–1400.

Farrell, B.H. and Twining-Ward, L. (2004) Reconceptualizing tourism. *Annals of Tourism Research* 31 (2), 274–295.

Faulkner, B. (2001) Towards a framework for tourism disaster management. *Tourism Management* 22 (2), 135–147.

Federal Emergency Management Agency (FEMA), United States Geological Survey (USGS), and the Pacific Disaster Center (PDC) (2017) *Hazus® Estimated Annualized Earthquake Losses for the United States*, FEMA P-366. Washington DC: Federal Emergency Management Agency.

Federal Reserve Bank of Kansas City (2016) How much economic damage do large earthquakes cause? [online]. See https://www.kansascityfed.org/publications/research/oke/articles/2016/economic-damage-large-earthquakes (accessed December 2019).

Folke, C. (2006) Resilience: The emergence of a perspective for social–ecological systems analyses. *Global Environmental Change* 16 (3), 253–267.

Folke, C., Carpenter, S., Walker, B., Scheffer, M., Chapin, T. and Rockström, J. (2010) Resilience thinking: Integrating resilience, adaptability and transformability. *Ecology and Society* 15 (4), 20.

Folke, C., Hahn, T., Olsson, P. and Norberg, J. (2005) Adaptive governance of social-ecological systems. *Annual Review of Environmental Resources* 30, 441–473.

Fountain, J. and Cradock-Henry, N. (2019) The road to recovery: Reimagining Kaikoura after a natural disaster. In G. Walters and J. Mair (eds) *Reputation and Image Recovery for the Tourism Industry*. Oxford: Goodfellow Publishers.

Foxworthy, B.L. and Hill, M. (1982) *Volcanic eruptions of 1980 at Mount St. Helens: The first 100 days*, USGS Professional Paper 1249. Washington DC: US Department of the Interior, Geological Survey.

Frazier, T.G., Walker, M.H., Kumari, A. and Thompson, C.M. (2013) Opportunities and constraints to hazard mitigation planning. *Applied Geography* 40, 52–60.

Fukui, M. and Ohe, Y. (2019) Assessing the role of social media in tourism recovery in tsunami-hit coastal areas in Tohoku, Japan. *Tourism Economics* in press. doi/full/10.1177/1354816618825014.

Geschwind, C.H. (2001) *California Earthquakes: Science, Risk, and the Politics of Hazard Mitigation*. Baltimore: John Hopkins University Press.

Ghimire, H.L. (2016) Tourism in Gorkha: A proposition to revive tourism after devastating earthquakes. *Journal of Tourism and Hospitality Education* 6, 67–94.

GNS Science, New Zealand (2019) Earthquakes: Understanding and monitoring seismic activity. [online]. See https://www.gns.cri.nz/Home/Our-Science/Natural-Hazards-and-Risks/Earthquakes (accessed December 2019).

Godschalk, D.R. (2003) Urban hazard mitigation: creating resilient cities. *Natural Hazards Review* 4 (3), 136–143.

Guo, Y., Zhang, J., Zhang, Y. and Zheng, C. (2018) Catalyst or barrier? The influence of place attachment on perceived community resilience in tourism destinations. *Sustainability* 10 (7), 2347.

Gurwitch, R.H., Kees, M., Becker, S.M., Schreiber, M., Pfefferbaum, B. and Diamond, D. (2004) When disaster strikes: Responding to the needs of children. *Prehospital and Disaster Medicine* 19 (1), 21–28.

Hall, C.M. (2002) Travel safety, terrorism and the media: The significance of the issue-attention cycle. *Current Issues in Tourism* 5 (5), 458–466.

Hall, C.M. (2005) *Tourism: Rethinking the Social Science of Mobility*. Harlow: Pearson.

Hall, C.M. (2008) *Tourism Planning* (2nd edn). Harlow: Pearson.

Hall, C.M. (2010) Crisis events in tourism: Subjects of crisis in tourism. *Current Issues in Tourism* 13 (5), 401–417.

Hall, C.M. (2012) New Zealand tourism recovering from earthquake: Local, national and international perspectives. Keynote address at the *2012 International Conference of Sport, Leisure and Hospitality Management Conference*, 11th May 2012–13th May 2012, Taipei.

Hall, C.M. (2014) *Tourism and Social Marketing*. Abingdon: Routledge.

Hall, C.M. and Amore, A. (2019) The 2015 Cricket World Cup in Christchurch. *Journal of Place Management and Development*. https://doi.org/10.1108/JPMD-04-2019-0029

Hall, C.M., Prayag, G. and Amore, A. (2018) *Tourism and Resilience: Individual, Organisational and Destination Perspectives*. Bristol: Channel View Publications.

Hall, C.M., Malinen, S., Vosslamber, R. and Wordsworth, R. (eds) (2016) *Business and Post-disaster Management: Business, Organisational and Consumer Resilience and the Christchurch Earthquakes*. Abingdon: Routledge.

Hasegawa, K. (2012) Facing nuclear risks: Lessons from the Fukushima nuclear disaster. *International Journal of Japanese Sociology* 21 (1), 84–91.

Henderson, J.C. (2013) The Great East Japan earthquake and tourism: A preliminary case study. *Tourism Recreation Research* 38 (1), 93–98.

Hernantes, J., Rich, E., Laugé, A., Labaka, L. and Sarriegi, J.M. (2013) Learning before the storm: Modeling multiple stakeholder activities in support of crisis management, a practical case. *Technological Forecasting and Social Change* 80 (9), 1742–1755.

Holling, C.S. and Gunderson, L.H. (2002) Resilienc and adaptive cycles. In L.H. Gunderson and C.S. Holling (eds) *Panarchy, Understanding the Transformations in Human and Natural Systems* (pp. 25–62). Washington, DC: Island Press.

Huan, T.C., Beaman, J. and Shelby, L. (2004) No-escape natural disaster: Mitigating impacts on tourism. *Annals of Tourism Research* 31 (2), 255–273.

Huang, J.H. and Min, J.C. (2002) Earthquake devastation and recovery in tourism: The Taiwan case. *Tourism Management* 23 (2), 145–154.

Huang, Y.C., Tseng, Y.P. and Petrick, J.F. (2008) Crisis management planning to restore tourism after disasters: A case study from Taiwan. *Journal of Travel & Tourism Marketing* 23 (2–4), 203–221.

Hystad, P.W. and Keller, P.C. (2008) Towards a destination tourism disaster management framework: Long-term lessons from a forest fire disaster. *Tourism Management* 29 (1), 151–162.

Insurance Information Institute (2019) *Facts + Statistics: Earthquakes and tsunamis*. New York: Insurance Information Institute. See https://www.iii.org/fact-statistic/facts-statistics-earthquakes-and-tsunamis#Earthquake%20Insurance,%202009-2018 (accessed December 2019).

Kamani-Fard, A., Hamdan Ahmad, M. and Remaz Ossen, D. (2012) The sense of place in the new homes of post-Bam earthquake reconstruction. *International Journal of Disaster Resilience in the Built Environment* 3 (3), 220–236.

Kelman, I., Spence, R., Palmer, J., Petal, M. and Saito, K. (2008) Tourists and disasters: Lessons from the 26 December 2004 tsunamis. *Journal of Coastal Conservation* 12 (3), 105–113.

Khazai, B., Mahdavian, F. and Platt, S. (2018) Tourism Recovery Scorecard (TOURS)– Benchmarking and monitoring progress on disaster recovery in tourism destinations. *International Journal of Disaster Risk Reduction* 27, 75–84.

Kimhi, S. (2016) Levels of resilience: Associations among individual, community, and national resilience. *Journal of Health Psychology* 21 (2), 164–170.

Laws, E. and Prideaux, B. (eds) (2005) *Tourism Crises: Management Responses and Theoretical Insight*. New York: The Haworth Hospitality Press.

Lew, A.A., Ng, P.T., Ni, C.C. and Wu, T.C. (2016) Community sustainability and resilience: Similarities, differences and indicators. *Tourism Geographies* 18 (1), 18–27.

Lin, Y., Kelemen, M. and Tresidder, R. (2018) Post-disaster tourism: Building resilience through community-led approaches in the aftermath of the 2011 disasters in Japan. *Journal of Sustainable Tourism* 26 (10), 1766–1783.

Lindell, M.K. and Prater, C.S. (2003) Assessing community impacts of natural disasters. *Natural Hazards Review* 4 (4), 176–185.

Linnenluecke, M.K. (2017) Resilience in business and management research: A review of influential publications and a research agenda. *International Journal of Management Reviews* 19 (1), 4–30.

Lowe, S.R., Manove, E.E. and Rhodes, J.E. (2013) Posttraumatic stress and posttraumatic growth among low-income mothers who survived Hurricane Katrina. *Journal of Consulting and Clinical Psychology* 81 (5), 877–889.

Luthans, F., Avey, J.B., Avolio, B.J., Norman, S.M. and Combs, G.M. (2006) Psychological capital development: Toward a micro-intervention. *Journal of Organizational Behavior* 27 (3), 387–393.

Mair, J., Ritchie, B.W. and Walters, G. (2016) Towards a research agenda for post-disaster and post-crisis recovery strategies for tourist destinations: A narrative review. *Current Issues in Tourism* 19 (1), 1–26.

Masterson, V.A., Stedman, R.C., Enqvist, J., Tengö, M., Giusti, M., Wahl, D. and Svedin, U. (2017) The contribution of sense of place to social-ecological systems research: A review and research agenda. *Ecology and Society* 22 (1). DOI: 10.5751/ES-08872-220149.

Mazzocchi, M. and Montini, A. (2001) Earthquake effects on tourism in central Italy. *Annals of Tourism Research* 28 (4), 1031–1046.

Mazzoni, S., Castori, G., Galasso, C., Calvi, P., Dreyer, R., Fischer, E., Fulco, A., Sorrentino, L., Wilson, J., Penna, A. and Magenes, G. (2018) 2016–2017 Central Italy Earthquake Sequence: Seismic retrofit policy and effectiveness. *Earthquake Spectra* 34 (4), 1671–1691.

Mendoza, C.A., Brida, J.G. and Garrido, N. (2012) The impact of earthquakes on Chile's international tourism demand. *Journal of Policy Research in Tourism, Leisure and Events* 4 (1), 48–60.

Mileti, D.S. and Passerini, E. (1996) A social explanation of urban relocation after earthquakes. *International Journal of Mass Emergencies and Disasters* 14 (1), 97–110.

Migoń, P. and Pijet-Migoń, E. (2019) Natural disasters, geotourism, and geo interpretation. *Geoheritage* 11 (2), 629–640.

Miller, F., Osbahr, H., Boyd, E., Thamalla, F., Bharwani, S., Ziervogel, G. and Nelson, D. (2010) Resilience and vulnerability: Complementary or conflicting concepts? *Ecology and Society* 15 (3), 11.

Morgan, J., Begg, A., Beaven, S., Schluter, P., Jamieson, K., Johal, S., Johnston, D. and Sparrow, M. (2015) Monitoring wellbeing during recovery from the 2010–2011 Canterbury earthquakes: The CERA Wellbeing Survey. *International Journal of Disaster Risk Reduction* 14, 96–103.

Moeller, S.D. (2006) 'Regarding the pain of others': Media, bias and the coverage of international disasters. *Journal of International Affairs* 59 (2), 173–196.

Muskat, B., Nakanishi, H. and Blackman, D. (2015) Integrating tourism into disaster recovery management: The case of the Great East Japan earthquake and tsunami 2011. In B. Ritchie and K. Campiranon (eds) *Tourism Crisis and Disaster Management in the Asia-Pacific* (pp. 97–115). Wallingford: CABI.

Mäntyniemi, P. (2012) An analysis of seismic risk from a tourism point of view. *Disasters* 36 (3), 465–476.

Nguyen, D.N., Imamura, F. and Iuchi, K. (2018) Barriers towards hotel disaster preparedness: Case studies of post 2011 Tsunami, Japan. *International Journal of Disaster Risk Reduction* 28, 585–594.

Oishi, S., Kimura, R., Hayashi, H., Tatsuki, S., Tamura, K., Ishii, K. and Tucker, J. (2015) Psychological adaptation to the Great Hanshin-Awaji Earthquake of 1995: 16 years later victims still report lower levels of subjective well-being. *Journal of Research in Personality* 55, 84–90.

Orchiston, C. (2013) Tourism business preparedness, resilience and disaster planning in a region of high seismic risk: The case of the Southern Alps, New Zealand. *Current Issues in Tourism* 16 (5), 477–494.

Orchiston, C. and Higham, J.E.S. (2016) Knowledge management and tourism recovery (de) marketing: The Christchurch earthquakes 2010–2011. *Current Issues in Tourism* 19 (1), 64–84.

Palm, R. (1998) Urban earthquake hazards: the impacts of culture on perceived risk and response in the USA and Japan' *Applied Geography* 18 (1), 35–46.

Paton, D. (2008) *Modelling Societal Resilience to Pandemic Hazards in Auckland*. GNS Science Report2008/13. Wellington: GNS Science.

Pike, S. (2004) *Destination Marketing Organisations*. Abingdon: Routledge.

Pottorff, S.M. and Neal, D.D.M. (1994) Marketing implications for post-disaster tourism destinations. *Journal of Travel & Tourism Marketing* 3 (1), 115–122.

Potter, S.H., Becker, J.S., Johnston, D.M. and Rossiter, K.P. (2015) An overview of the impacts of the 2010-2011 Canterbury earthquakes. *International Journal of Disaster Risk Reduction* 14, 6–14.

Pizzo, B. (2015) Problematizing resilience: Implications for planning theory and practice. *Cities* 43, 133–140.

Prayag, G. (2018) Symbiotic relationship or not? Understanding resilience and crisis management in tourism. *Tourism Management Perspectives* 25, 133–135.

Prayag, G., Chowdhury, M., Spector, S. and Orchiston, C. (2018) Organizational resilience and financial performance. *Annals of Tourism Research* 73 (C), 193–196.

Prayag, G., Fieger, P. and Rice, J. (2019a) Tourism expenditure in post-earthquake Christchurch, New Zealand. *Anatolia* 30 (1), 47–60.

Prayag, G., Spector, S., Orchiston, C. and Chowdhury, M. (2019b) Psychological resilience, organizational resilience and life satisfaction in tourism firms: Insights from the Canterbury earthquakes. *Current Issues in Tourism* [online]. DOI: 10.1080/13683500.2019.1607832.

Price, L.L., Coulter, R.A., Strizhakova, Y. and Schultz, A.E. (2018) The fresh start mindset: Transforming consumers' lives. *Journal of Consumer Research* 45 (1), 21–48.

Prideaux, B., Coghlan, A. and Falco-Mammone, F. (2008) Post crisis recovery: The case of after cyclone Larry. *Journal of Travel & Tourism Marketing* 23 (2–4), 163–174.

Rangel, L.E. and Lévêque, F. (2014) How Fukushima Dai-ichi core meltdown changed the probability of nuclear accidents?. *Safety Science* 64, 90–98.

Rehdanz, K., Welsch, H., Narita, D. and Okubo, T. (2015) Well-being effects of a major natural disaster: The case of Fukushima. *Journal of Economic Behavior & Organization* 116, 500–517.

Reich, J.W. (2006) Three psychological principles of resilience in natural disasters. *Disaster Prevention and Management* 15 (5), 793–798.

Ritchie, B. (2008) Tourism disaster planning and management: From response and recovery to reduction and readiness. *Current Issues in Tourism* 11 (4), 315–348.

Ritchie, B.W. (2009) *Crisis and Disaster Management for Tourism*. Bristol: Channel View Publications.

Rogers, P., Burnside-Lawry, J., Dragisic, J. and Mills, C. (2016) Collaboration and communication: Building a research agenda and way of working towards community disaster resilience. *Disaster Prevention and Management* 25 (1), 75–90.

Rose, A. (2011) Resilience and sustainability in the face of disasters. *Environmental Innovation and Societal Transitions* 1 (1), 96–100.

Russell, L.A., Goltz, J.D. and Bourque, L.B. (1995) Preparedness and hazard mitigation actions before and after two earthquakes. *Environment and Behavior* 27 (6), 744–770.

Sengezer, B. and Koç, E. (2005) A critical analysis of earthquakes and urban planning in Turkey. *Disasters* 29 (2), 171–194.

Seville, E. (2018) Building resilience: How to have a positive impact at the organizational and individual employee level. *Development and Learning in Organizations: An International Journal* 32 (3), 15–18.

Seyfi, S. and Hall, C.M. (2020) *Tourism, Sanctions and Boycotts*. Abingdon: Routledge.

Sherrieb, K., Norris, F.H. and Galea, S. (2010) Measuring capacities for community resilience. *Social Indicators Research* 99 (2), 227–247.

Sönmez, S.F., Apostolopoulos, Y. and Tarlow, P. (1999) Tourism in crisis: Managing the effects of terrorism. *Journal of Travel Research* 38 (1), 13–18.

Steckler, M.S., Stein, S., Akhter, S.H. and Seeber, L. (2018) The wicked problem of earthquake hazard in developing countries. *Eos* 99. https://doi.org/10.1029/2018EO093625.

Subedi, S., Sharma, G.N., Dahal, S., Banjara, M.R. and Pandey, B.D. (2018) The health sector response to the 2015 earthquake in Nepal. *Disaster Medicine and Public Health Preparedness* 12 (4), 543–547.

Sugano, S. (2016) The well-being of elderly survivors after natural disasters: Measuring the impact of the Great East Japan Earthquake. *The Japanese Economic Review* 67 (2), 211–229.

Sutcliffe, K.M. and Vogus, T. (2003) Organizing for resilience. In K.S. Cameron, J.E. Dutton and R.E. Quinn (eds) *Positive Organizational Scholarship: Foundations of a New Discipline* (pp. 94–110). San Francisco, CA: Berrett-Koehler.

Tang, Y. (2014) Travel motivation, destination image and visitor satisfaction of international tourists after the 2008 Wenchuan earthquake: A structural modelling approach. *Asia Pacific Journal of Tourism Research* 19 (11), 1260–1277.

Tao, B.C., Goh, E., Huang, S. and Moyle, B. (2019) Travel constraint perceptions of people with mobility disability: A study of Sichuan earthquake survivors. *Tourism Recreation Research* 44 (2), 203–216.

Thompson, M.A., Owen, S., Lindsay, J.M., Leonard, G.S. and Cronin, S.J. (2017) Scientist and stakeholder perspectives of transdisciplinary research: Early attitudes, expectations, and tensions. *Environmental Science & Policy* 74, 30–39.

Thoresen, S., Tønnessen, A., Lindgaard, C.V., Andreassen, A.L. and Weisæth, L. (2009) Stressful but rewarding: Norwegian personnel mobilised for the 2004 tsunami disaster. *Disasters* 33 (3), 353–368.

Tierney, K. and Oliver-Smith, A. (2012) Social dimensions of disaster recovery. *International Journal of Mass Emergencies & Disasters* 30 (2), 123–146.

Tsai, C.H. and Chen, C.W. (2010) An earthquake disaster management mechanism based on risk assessment information for the tourism industry-a case study from the island of Taiwan. *Tourism Management* 31 (4), 470–481.

Tucker, H., Shelton, E.J. and Bae, H. (2017) Post-disaster tourism: Towards a tourism of transition. *Tourist Studies* 17 (3), 306–327.

United Nations Office for Disaster Risk Reduction (UNISDR) (2015) *Proposed updated terminology on disaster risk reduction: A technical review background paper*. See http://www.preventionweb.net/files/45462_backgoundpaperonterminologyaugust20.pdf (accessed September 2016).

United States Geological Survey (USGS) (2019) *The Science of Earthquakes*. [online]. See https://www.usgs.gov/natural-hazards/earthquake-hazards/science/science-earthquakes?qt-science_center_objects=0#qt-science_center_objects (accessed December 2019).

Van der Voort, N. and Vanclay, F. (2015) Social impacts of earthquakes caused by gas extraction in the Province of Groningen, The Netherlands. *Environmental Impact Assessment Review* 50, 1–15.

Veer, E., Ozanne, L. and Hall, C.M. (2016) Sharing cathartic stories online: The internet as a means of expression following a crisis event. *Journal of Consumer Behaviour* 15 (4), 314–324

Walters, G. and Clulow, V. (2010) The tourism market's response to the 2009 Black Saturday bushfires: The case of Gippsland. *Journal of Travel & Tourism Marketing* 27 (8), 844–857.

Wang, S., Chen, S. and Xu, H. (2019) Resident attitudes towards dark tourism, a perspective of place-based identity motives. *Current Issues in Tourism* 22 (13), 1601–1616.

Wang, Y.S. (2009) The impact of crisis events and macroeconomic activity on Taiwan's international inbound tourism demand. *Tourism Management* 30 (1), 75–82.

Wearing, S., Beirman, D. and Grabowski, S. (2020) Engaging volunteer tourism in post-disaster recovery in Nepal. *Annals of Tourism Research* 80, 102802.

Whitman, Z., Stevenson, J., Kachali, H., Seville, E., Vargo, J. and Wilson, T. (2014) Organisational resilience following the Darfield earthquake of 2010. *Disasters* 38 (1), 148–177.

Whitman, Z.R., Wilson, T.M., Seville, E., Vargo, J., Stevenson, J.R., Kachali, H. and Cole, J. (2013) Rural organizational impacts, mitigation strategies, and resilience to the 2010 Darfield earthquake, New Zealand. *Natural Hazards* 69 (3), 1849–1875.

Wright, D. and Sharpley, R. (2018) Local community perceptions of disaster tourism: The case of L'Aquila, Italy. *Current Issues in Tourism* 21 (14), 1569–1585.

Wu, J.Y. and Lindell, M.K. (2004) Housing reconstruction after two major earthquakes: The 1994 Northridge earthquake in the United States and the 1999 Chi-Chi earthquake in Taiwan. *Disasters* 28 (1), 63–81.

Wu, L. and Hayashi, H. (2014) The impact of disasters on Japan's inbound tourism demand. *Journal of Disaster Research* 9, 699–708.

Wu, L. and Walters, G. (2016) Chinese travel behavior in response to disastrous events: The case of the Japan earthquake. *Journal of China Tourism Research* 12 (2), 216–231.

Yan, B.J., Zhang, J., Zhang, H.L., Lu, S.J. and Guo, Y.R. (2016) Investigating the motivation–experience relationship in a dark tourism space: A case study of the Beichuan earthquake relics, China. *Tourism Management* 53, 108–121.

Yi, H. and Yang, J. (2014) Research trends of post disaster reconstruction: The past and the future. *Habitat International* 42, 21–29.

2 The Resilience of a Tourist Destination: Seismic Risk Perception by Tourism Operators in the Etna Area, Italy

Barbara Martini and Marco Platania

Introduction

The concepts of social ecological resilience and vulnerability are gaining increasing interest in tourism (Hall *et al.*, 2017). Tourism enterprises are a core component of community resilience in destination regions. They are, especially in some areas, not only an important part of the business sector but also play an important role in cases of disaster. This chapter focuses on vulnerability and resilience of tourism enterprises at the municipal scale in order to develop policies for building resilient communities with a specific capacity for disaster preparedness and enhancing the sector's ability to recover from crises and disasters.

The growing interest in resilience and vulnerability in tourism has arguably been stimulated by the crises and disasters that have affected the sector since 2000 including the New York twin towers attack in 2001 and Hurricane Katrina in 2005. Therefore, it has become increasingly important to understand and manage the resilience of vulnerable socio-economic sectors such as tourism. In order to take into consideration not only the economic components of a system but also the social ecological dimensions we will use the approach based on *dynamics of complex, adaptive, social-ecological systems* (SES) – the ability of groups or communities to cope with external stresses and disturbances as a result of social, political and environmental change (Holling, 1973, 1996, 2001; Gunderson, 2000; Adger, 2000; Walker *et al.*, 2004; Carpenter *et al.*, 2005) in which social and ecological resilience are inter-related to understand the relationship between vulnerability and resilience of tourism enterprises. Following this approach, resilience can be defined as the ability of groups or communities to cope with external stressors and disturbances, while maintaining their functional characteristics and defined identity (Adger, 2000). Furthermore, resilience is a process (Pendall *et al.*, 2010) and it includes those inherent conditions that allow the system to absorb impacts and disturbance and cope with an event, as well as post-event situations. Finally, it is an adaptive process that facilitates the ability of the social system to re-organize, change, and learn in response to a threat (Cutter *et al.*, 2008) and reorganize while undergoing change so as to still retain essentially the same function, structure, identity, and feedbacks (Walker & Meyers, 2004).

The definition of vulnerability varies within and across research traditions (Akter & Mallick, 2013). The United Nations International Strategy for Disaster Reduction (UNISDR, 2004: 16) defines vulnerability as 'The conditions determined by physical, social, economic and environmental factors or processes, which increase the susceptibility of a community to the impact of hazards'. In contrast, the United National Development Programme (UNDP, 2004: 98) defines vulnerability as 'a human condition or process resulting from physical, social, economic and environmental factors, which determine the likelihood and scale of damage from the impact of a given hazard'. The disaster risk literature focuses on the propensity of exposed elements to suffer adverse effects when impacted by a hazard (Cannon, 1994, 2006; Jansen *et al.*, 2006). The human geography approach, by contrast, considers vulnerability a combination of sensitivity, exposure and response capacity (Adger, 2006; Gallopin, 2006). Sensitivity is seen as an internal property of the system and refers to the degree to which a system is likely to be affected by an internal or external disturbance (Gallopin, 2006). Exposure refers to the degree, duration and/or extent to which the system is in contact with or subject to a disturbance (Gallopin, 2006). The response capacity is the system's ability to respond to or cope with the disturbance. The aim of this chapter is to analyse how people and societies respond to environmental hazard, identify what factors influence their choice of adjustment and determine how to mitigate the risk and impact of environmental hazard (Cutter, 1996). Vulnerability, in this case, can be considered not only as exposure but also as a social condition (a measure of resilience to hazards) and, finally, 'the integration of potential exposures and societal resilience with a specific focus on places or regions' (Cutter *et al.*, 2003: 243).

Vulnerability is multi-dimensional and differential (varies across physical space and among and within social groups), scale dependent (with regard to time, space and units of analysis such as individual, household, region, system) and dynamic (the characteristics and driving forces of vulnerability change over time) (Vogel & O'Brien, 2004). Furthermore, social vulnerability is much more than the likelihood of buildings collapsing and infrastructure being damaged but is a set of characteristics that includes a person's initial well-being; livelihood and resilience (assets and capitals, income and exchange options, qualifications); self-protection (the degree of protection afforded by capability and willingness to build a safe home, use a safe site); social protection (forms of hazard preparedness provided by society more generally, e.g. building codes, mitigation measures, shelters, preparedness); and social and political networks and institutions (social capital, the role of the institutional environment in setting good conditions for hazard precaution, peoples' rights to express needs and of access to preparedness) (Cannon et al., 2003). Following this approach, vulnerability is not only partially determined by the type of hazard but it also depends on social and institutional components. The relationship between resilience and vulnerability is still unclear in the literature. Following Cutter and Emrich (2006), vulnerability is the pre-event, inherent characteristics or qualities of social systems that create the potential for harm. Vulnerability is a function of the exposure and sensitivity of system (Adger, 2006; Cutter, 1996). Resilience, by contrast, is the ability of a social system to respond and recover from disasters and includes those inherent conditions that allow the system to absorb impacts and cope with an event, as well as post-event, adaptive processes that facilitate the ability of the social system to re-organize, change and learn in response to a threat. Different levels of vulnerability and resilience can generate disparities among social systems.

Social vulnerability is 'the differential exposure to stresses experienced or anticipated by the different units exposed', a dynamic process 'rooted in the actions and multiple attributes of human actors' often determined by social networks in social, economic, political and environmental interactions 'manifested simultaneously on more than one scale' influenced and driven by multiple stresses (Cutter et al., 2003: 243). Consequently, the concept of social vulnerability refers to more than socioeconomic impacts, since it can also encompass features of potential physical damage in the built environment (Cutter et al., 2003). Social resilience is about the abilities of social entities to tolerate, absorb, cope with and adjust to threats (Adger, 2000). Social resilience is composed of several interrelated dimensions (Voss, 2008; Lorenz, 2010; Armitage et al., 2012; Keck & Sakdapolrak, 2013) and it can refer to individuals (Butler et al., 2007; Williams & Drury, 2009), communities (Norris et al., 2008) and society as a whole (Adger, 2000; Hall et al., 2017). Social capital is important not only to cope with adverse shock but also for reduction of disaster risks and recovery from disaster shock. Social capital is embedded in social relationships (Bourdieu, 1983; Coleman, 1988; Putnam, 1993), nevertheless Coleman (1988) focuses on the individual use, Bourdieu (1983) on the use of social capital by certain social groups and Putnam (1993) focuses on the function of social capital for communities.

The aim of the chapter is threefold. First, analysing the degree of preparedness and the risk perceptions of disaster of tourism operators in the Etna region. Second, to evaluate community resilience in a tourism destination. Finally, investigate the level of trust of tourism operators with respect to government and institutions.

Study Area

The analysis is focused near Catania, the second largest city of Sicily located on the east coast of Sicily on the Ionian Sea. Catania is a metropolitan area and one of the largest cities in Italy in terms of inhabitants. This area has been growing in terms of tourism in recent years and one of the most interesting tourist attractions is the Etna volcano.

Mount Etna is not only the highest active volcano in Europe (3329 m) but it is also one of the most active volcanoes in the world. It is the highest peak in Italian south of the Alps and covers an area of 1190 km^2 (459 sq mi) with a basal circumference of 140 km. Lying above the convergent plate margin between the African and Eurasian Plates, Mount Etna is an active volcano and sometimes it switches to an eruptive phase but its eruptions rarely endanger inhabitants because the lava flows move slowly. These slow movements give the opportunity to evacuate people before there is any damage. However, despite this, Etna's lava flows have a high destructive power. They can destroy not only houses, agriculture and roads but also tourist facilities. Furthermore, the eruptions often produce lava dust which generates substantial discomfort for daily life. Nevertheless, there are also same advantages related to the volcanic soil that can support agriculture in vineyards and orchards.

In order to protect this habitat, which is characterized by high natural and cultural values, in 1987 Sicily Region established a protected area, the Etna Park. From an administrative point of view, the area's governance is complex due the fact that all the municipalities located around the Etna have administrative territories that reach the summit of Etna. This governance has proved to be a problem for land management. In June 2013, the Etna Park was added to the list of UNESCO World Heritage sites. Nevertheless, the Etna region provides substantial risk not only because of the volcano but also because it is located in an area characterized by earthquakes (Gruppo di Lavoro MPS, 2004). The research on which this chapter is based will take into consideration the hypothesis that an earthquake will occur in this area.

Methodology and Sampling

The research aim is to investigate the preparedness and the risk perception of tourism firms in an area characterized by seismic risk. These elements are not only related to community resilience (Cutter *et al.*, 2008; Norris *et al.*, 2008) but also to vulnerability. Social resilience can be defined as the ability of a social actor to cope with and overcome the adversity, moreover, it takes into consideration adaptive capacity, their ability to learn from past experiences and adjust themselves to future challenges in their everyday lives and their transformative capacities, their ability to craft sets of institutions that foster individual welfare and sustainable societal robustness towards future crises.

In order to explore social resilience, vulnerability and informal networks, a questionnaire was used that was elaborated from previous studies applied in New Zealand (Orchiston, 2010, 2012, 2013). The survey was divided into four sections. The first one regards the risk perception in order to investigate the threat awareness with respect to an earthquake. It takes into consideration not only hypothetical damage but also previous experience in coping with earthquake in accordance with previously defined adaptive capacity and vulnerability. Knowledge acquired during the past can be useful and it can influence the risk perceptions and the preparedness level in case an earthquake occurs again. The second section explores the degree of personal and community preparedness if a seismic event was to occur. This section provides information regarding the role of the community and the sense of belonging, in terms of involvement in preparatory activities to cope with a natural shock and the capacity to manage vulnerability. The aim of this section is to capture social resilience and to investigate the intangible relationships that contribute to the creation of social capital (Putnam, 1993). The third part of the survey investigates the degree of business preparedness and provides information regarding the tourism firm's capacity to cope with shock. The final section provides information on the socioeconomic characteristics (age, level of education, social situation and so on) of the interviewees.

The questionnaire administration was preceded by an explanation on the characteristics of an earthquake with a 6 to 8 magnitude, taking as a reference the 2016 seismic phenomenon in the centre of Italy in 2016 (Amatrice/Central Italy earthquakes). With respect to this hypothetical event, the interview asked how the tourism industry would react taking into consideration not only the hypothetical damages to economic activity, but also impacts on people, communities and cities. The survey was undertaken through face-to-face interview with 60 people belonging to 11 different municipalities in the Etna area. The survey was conducted during the months of November 2016 and April 2017.

The respondents were owners or managers of a tourism business. In the Etna area the hosting of visitors is undertaken by hotels, bed and breakfasts and agritourism operators. The sample consists of representatives of 49 hotels and 11 other host accommodation providers. In a socioecological tourism system framework, these other accommodation operators, cheaper and less equipped in the case of

Table 2.1 Socioeconomic characteristics of respondents

	n	%
Age		
Less than 20	1	1.7
21–40	27	45.0
41–64	29	48.3
65 or above	3	5.0
Gender		
male	37	61.7
female	23	38.3
Household		
Family without children	12	20.0
Family with children	28	46.7
other	4	6.7
Living alone	10	16.7
Living with others	6	10.0
Highest qualification level		
High school diploma	25	41.7
3 year degree	14	23.3
5 year degree	21	35.0
Role in business		
Head of reception	12	20.0
Manager	25	41.7
Owner	23	38.3

natural disaster, must be taken into consideration because they represent an important fraction of tourist supply.

The social and demographic characteristics of interviewees are summarized in Table 2.1. The interviewees are mainly male (61.7%), middle age (41–64: 48.3%), have a family with children (46.7%) with the highest qualification being a high school diploma (41.7%).

Results

Risk perceptions

The starting point was to analyse the perception regarding the relative dangerousness of an earthquake in comparison with other natural hazards that can occur in the same area, including volcanic eruption, earthquake, drought, storm or flood. Respondents were asked to indicate on a scale from 1 (very unlikely) to 5 (very likely) the potential impact on livelihood.

The potential impact on livelihood from an earthquake has been considered likely or very likely by 82% of the respondents, with only 1.7%

Table 2.2 Provisional timeframes of a fault earthquake in the Etna area

	n	%
Within the next year	1	1.7
Next 10 years	3	5.0
Next 30 years	8	13.3
Next 50 years	6	10.0
More than 50 years	42	70.0
Total	60	100.0

answering unlikely or very unlikely. Storms and droughts were considered likely or very likely by 11.7% of respondents. Volcano eruptions were thought to be very unlikely by 40% of those interviewed. Respondents considered an earthquake a catastrophic event and have the perception that not only it is dangerous but also the degree of risk associated with this event is very high. The second question asked interviewees the probability that 'an earthquake of magnitude 8 will occur in the future in the Etna area'. The results are summarized in Table 2.2.

The interviewees perceive that an earthquake is dangerous and it can potentially be destructive, although they assign a low probability to the event occurring in the immediate future. The majority of interviewees have a clear perception of the risk because they live in a seismic area, nevertheless, they have the idea that the event will happen far in the future with 70% of respondents believing that an earthquake will happen in more the 50 years. As a consequence, tourism operators do not consider the earthquake as an immediate threat. This increases the level of vulnerability. As previously underlined, vulnerability is not only related to the hazard but also depends on the social and the institutional components. In this case the risk perception and the preparedness level are low and this increases the level of vulnerability.

The second question aimed to investigate the damage that can occur after an earthquake of magnitude 8. Each choice (damage to personal safety, ownerships, tourism business and community) could be ranked as: not at all, possible, not sure, probable and very probable. The interviewees considered damage to personal safety, ownerships, tourism business and community very probable (Table 2.3). The earthquake is perceived as a

Table 2.3 Level of threat due to earthquake (% value)

	Not at all	Possible	Not sure	Probable	Very probable	Total
Personal safety	10.0	5.0	1.7	18.3	65.0	100.0
Ownership	5.0	5.0	6.7	25.0	58.3	100.0
Tourism business	3.3	1.7	5.0	20.0	70.0	100.0
Community	3.3	1.7	5.0	18.3	71.7	100.0

catastrophic event, very dangerous not only for ownerships and tourism business but also for personal safety and for community safety.

With respect to a natural disaster and in order to create resilient places and communities the level of preparedness of individuals and the community is important. In order to explore these components respondents were asked to indicate on a Likert scale their agreement with respect to four statements: 1. a fault earthquake will be too destructive to bother preparing for; 2. it is unnecessary to prepare for an earthquake because assistance will be provided by the government; 3. getting prepared for an earthquake will significantly reduce the damage to my home; and 4. getting prepared for an earthquake will help my business to recover after the event. The results are summarized in Figure 2.1. The majority of interviewees (57%) are strongly persuaded that an earthquake in the Etna area will be so destructive that to be prepared is unnecessary. Only 28% disagreed or strongly disagreed with this statement. This answer highlights the idea that 'if it happens, it will happen; we can't do anything'. This kind of approach underlines a mix between fatalism and resignation. Moreover, the results show a strong distrust with respect to institutions with 62.2% of respondents not believing that assistance will be provided by regional government. This behaviour not only increases the level of vulnerability but also decrease resilience. Moreover, the preparedness level can reduce damage to houses

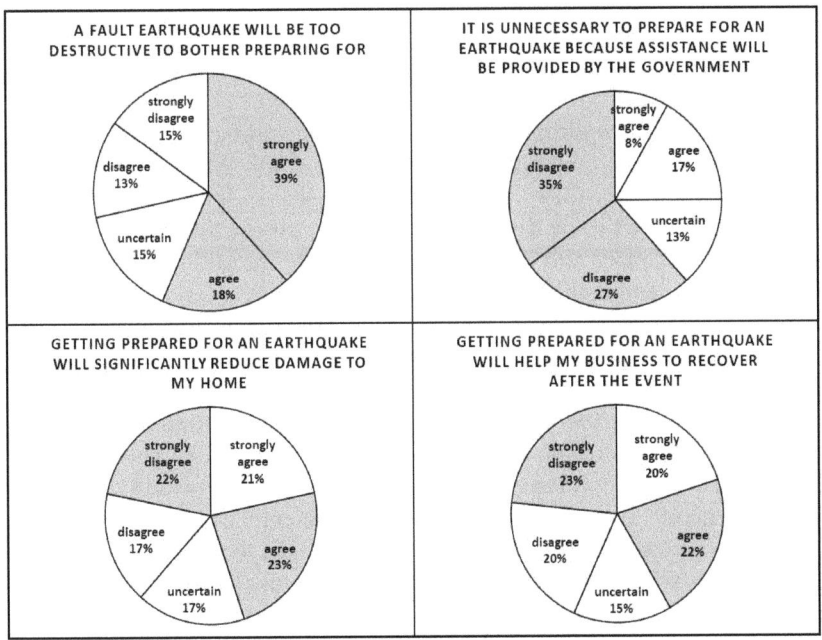

Figure 2.1 Fault earthquake risk: Perception of preparedness

Table 2.4 Earthquake risk: Sources of information

	n	%
I have never sought information	13	21.7
Through internet	37	61.7
Through civil protection agencies	11	18.3
Through local council (municipality)	6	10.0
Through regional council	3	5.0

and business for 50% of interviewees. Results from these questions suggest that a significant proportion of tourism operators retain the notion that the earthquake will hit so strongly that nothing can be done to avoid it. They are not able to figure out the consequences and how to deal with them.

Personal and community preparedness

Four prominent sources of information regarding earthquake risks were listed. As shown in Table 2.4, 22% of respondents had never sought information regarding earthquake. The remaining interviewees use the internet (62%), civil protection (18%), Local Councils (10%) and the Regional Council (5%) as information sources. These results highlight a lack of trust in institutions given that they prefer to be informed via the internet than from appointed authorities. In order to increase the level of preparedness institutions therefore need to create a more trustworthy environment in order to encourage industry to contact them in terms of help.

Resilience is a multidimensional process including not only the individual but also the community (Cutter et al., 2008; Magis, 2010; Adger, 2000; Norris et al., 2008). Moreover, resilience it is not a fixed place attribute but needs to be developed by taking into consideration the social components of place and industry processes. The questionnaire underlines a lack of trust and communications among different social components of a given area. Being part of a community means also improving the degree of engagement of each member. The greater the level of involvement the better the ability to react and to cope with a disaster. In order to take into consideration these elements into the analysis the questionnaire includes some questions regarding the degree of risk preparedness. The first question regards the community preparedness initiatives, including community emergency response planning and the dissemination of information regarding hazards involving citizens (Table 2.5). Forty-five percent of respondents state that their community has an emergency response plan but only 21.7% has received information from local or regional authorities concerning earthquake hazard. Sixty-five percent of respondents had also never been involved in any emergency preparedness meetings

Table 2.5 Community preparedness initiatives

	%
Does your community have an emergency response plan?	45.0
Did you receive information about the dangers of earthquakes by local or regional authorities?	21.7
Have you discussed the danger of an earthquake with your family or your neighbours?	66.7
You or some family members have been involved in emergency preparedness meetings, run by your local council or civil defense or schools.	35.0

Table 2.6 Respondent perception of preparedness with respect to a seismic event (% value)

	Not all prepared	2	3	4	Very prepared	Total
Yourself	35.0	38.3	20.0	1.7	5.0	100.0
Your community	45.0	35.0	15.0	5.0	0.0	100.0
Your business	30.0	28.3	31.7	8.3	1.7	100.0
Your municipality	36.7	31.7	25.0	6.7	0.0	100.0
Your regional government	43.3	28.3	18.3	6.7	3.3	100.0

although 66.7% had discussed earthquake preparedness with family/household, or neighbours.

The second question regarded the perception of respondents with respect to the degree of preparation for a seismic event. The results are show in Table 2.6. The interviewees were asked to indicate on a Likert scale their perception of the level of preparedness of themselves, their community, tourism businesses, the local council and the regional government. The main finding is that the perception, in terms of the level of preparedness, at different scales (from individual to community to institution) is very low. As a consequence, the tourism operators working in Etna area appear unprepared for this event. Moreover, the perception of being unprepared it is not only at an individual level but also at the community and business levels. Finally, the respondents believe that the local authorities are also unprepared.

Five statements regarding perceptions of community belonging were designed to explore the sense of community. The latter is an important dimension both of resilience and vulnerability. The results are shown in Table 2.7. More than the 50% of tourism operators do not want to move out of their community. They feel a part of the community (63.3%) and are planning to remain in the community for many years (61.6%) as well as feeling loyal to the people in their community (86.7%). The results show that social networks and the sense of belonging to their community are very strong. Yet, despite this, the level of trustiness in institutions is very low.

Table 2.7 Perceptions of community belonging (% value)

	Totally agree	2	3	4	Totally disagree	Total
Given the opportunity, I would like to move out of this community	16.7	21.7	15.0	30.0	16.7	100.0
I believe my neighbours will help me in an emergency	13.3	45.0	26.7	10.0	5.0	100.0
I feel I belong to my community	10.0	53.3	23.3	10.0	3.3	100.0
I feel loyal to the people in my community	36.7	50.0	8.3	0.0	5.0	100.0
I plan to remain a resident of this community for many years	33.3	28.3	25.0	5.0	8.3	100.0

Business preparedness

The level of preparedness plays an important role in coping with and recovering from a natural disaster and is an important component of resilience and vulnerability. Regarding the business preparation, the results show that the 51.7% have a training program for new staff members, and that 50% had specific training for emergencies but only a third provide specific training devoted to natural disaster management (Table 2.8).

The level of preparedness is also influenced by the level of involvement with respect to the problem of earthquakes. The frequent discussion of a topic, with the family or with a partner, is a way to increase the level of sensitivity with respect to the hazard. Only 7% of respondents had discussed the topic with customers at least once a week or more and only the 2% talked about it together with other business owners (Figure 2.2). Most of the respondents never or rarely talk about earthquakes in the business environment.

Tourism operators were asked seven questions regarding the response and recovery phases of a major earthquake in order to explore their opinion regarding the severity of consequences and the potential impact on their activities (Figures 2.3 and 2.4). A five-point Likert scale from strongly agree to strongly disagree was provided. There is a perception that an earthquake will severely impact on the business

Table 2.8 Preparation acts for business resilience

	n	%
Do you have a training program for new staff?	31	51.7
If yes, does it include a section on dealing with natural disasters?	20	33.3
Do you run evacuation drills?	27	45.0
Do you have on-going staff training about emergencies	50	50.0

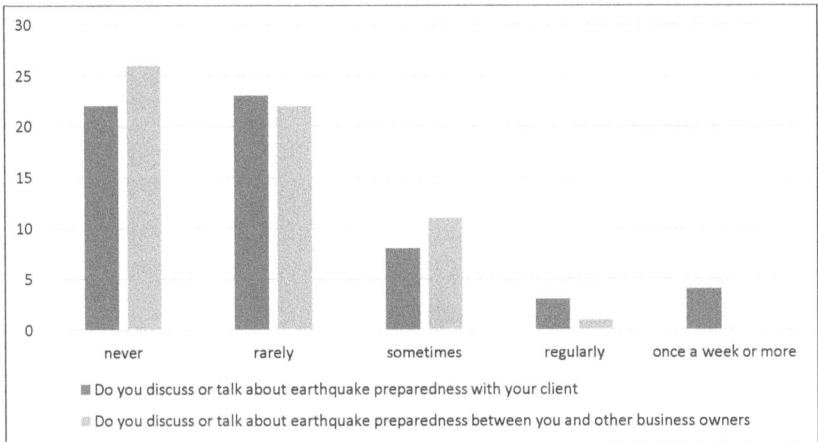

Figure 2.2 Discussion on earthquake preparedness (% value)

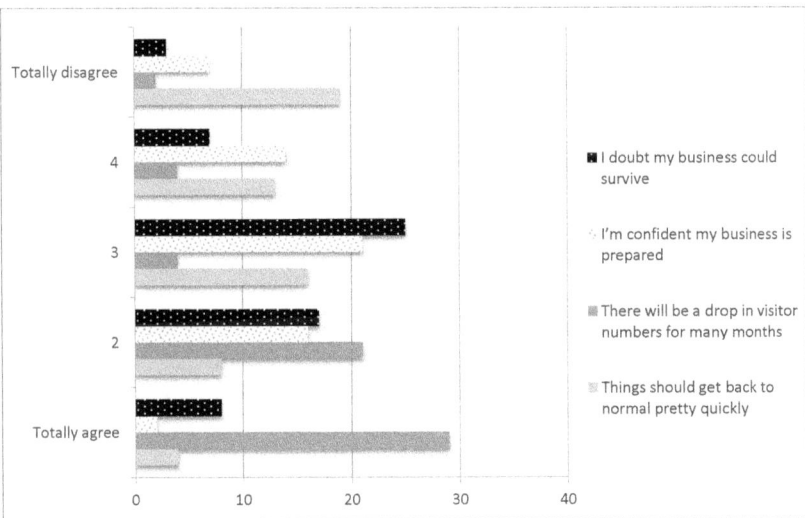

Figure 2.3 Opinion regarding consequences of fault earthquake (% value)

and the number of tourists while respondents are also expecting a long recovery phase because they do not expect a prompt return to 'normal' (Figure 2.3). Nevertheless, more than half of tourism operators either agree or totally agree that the business community would support each other after an earthquake during the recovery period (Figure 2.4). Moreover, they also feel responsible for their staff. Once again, the results show a very high degree of social networking at the individual and community levels.

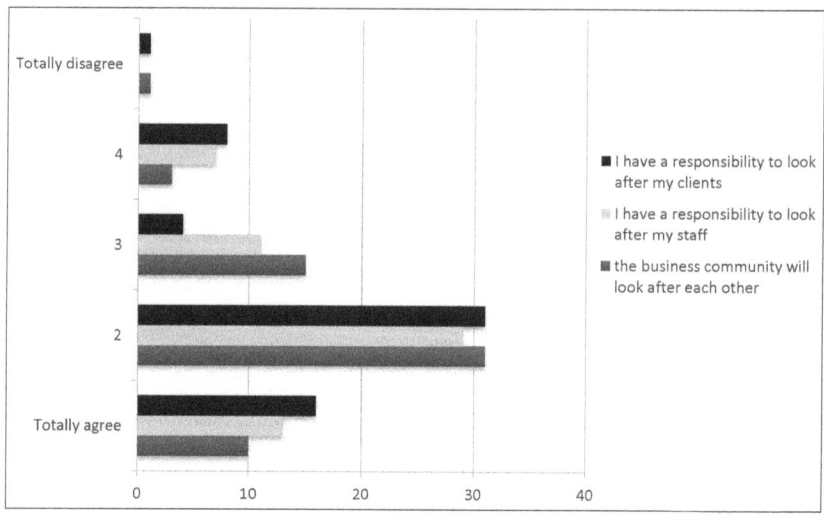

Figure 2.4 Opinion regarding consequences of fault earthquake (% value)

Conclusions: Resilience, Vulnerability and Tourism in the Etna Region

Tourism is a social and economic activity that is integral to destination communities. As a result, tourism must be taken into consideration by destinations in order to increase resilience and decrease vulnerability. Resilience and vulnerability embody important social components i.e. social networks and level of preparedness with respect to natural hazards, that need to be considered in order to create resilient places and to develop appropriate destination policies and actions.

Nevertheless, social components are very often intangible and it can be difficult to capture them. In order to explore these important dimensions, this analysis surveyed members of the tourism industry in the Etna region that provide accommodation to visitors. The results show that the area has some strengths but also some weaknesses. Among the positive points there is a very strong sense of community, sense of belonging and social networks. Overall, the amount of social capital appears high and the degree of trust within the community is very high. Despite this, the level of trust in institutions is very low. Moreover, the perception of earthquake risk is very low and, as a consequence, the level of preparedness is also very low. Furthermore, interviewees appear to believe that a natural disaster can be so disruptive that nothing can be done and everything will be damaged. As a result of these various factors, the majority of respondents are not willing to prepare for a major earthquake event.

The results underline a critical situation in which it is difficult to construct policies that will increase resilience and reducing vulnerability.

However, the study highlights that to improve the level of preparation for earthquakes communities must be guided by good local leadership in which trust has been developed between the various stakeholders. The communities studied are very critical of the level of regional support and this generates problems of trust. In order to respond, local plans must be made in conjunction with the local population in order to generate trust and give effect to plans and policies. Finally, it becomes clear that the stronger the sense of community, the better will be its ability to transmit information, to develop trust in institutions and to increase the sense of belonging, all fundamental elements both before and after the earthquake.

References

Adger, N. (2006) Vulnerability. *Global Environmental Change* 16 (3), 268–281.
Adger, W.N. (2000) 'Social and ecological resilience: Are they related?' *Progress in Human Geography* 24 (3), 347–364.
Armitage, D., Béné, C., Charles, A.T., Johnson, D. and Allison, E.H. (2012) The interplay of well-being and resilience in applying a social-ecological perspective. *Ecology and Society* 17 (4), 15.
Akter, S. and Mallick, B. (2013) The poverty–vulnerability–resilience nexus: Evidence from Bangladesh. *Ecological Economics* 96, 114–124.
Bourdieu, P. (1983) The forms of capital. In J.G. Richardson (ed.) *Handbook of Theory and Research for Sociology and Education* (pp. 41–58). New York: Greenwood Press.
Butler, D., Morland, L.A. and Leskin, G.A. (2007) Psychological resilience in the face of terrorism. In B. Bongar, L.M. Brown, L.E. Beutler, J.N. Breckenridge and P.G. Zimbardo (eds) *Psychology of Terrorism* (pp. 400–417). Oxford: Oxford University Press.
Cannon, T. (1994) Vulnerability analysis and the explanation of natural disaster. In S. Varley (ed.) *Disasters, Development and Environment* (pp. 13–29). Chichester: Wiley-Blackwell.
Cannon, T. (2006) Vulnerability analysis, livelihoods and disaster. In W. Amman, S. Dennenmann and L. Vuliet (eds) *Risk 21 – Coping with Risks Due to Natural Hazards in the 21st Century* (pp. 41–50). *Proceedings of the RISK21 Workshop, Monte Verità, Ascona, Switzerland, 28 November – 3 December.* Abingdon: Taylor Francis.
Cannon, T., Twigg, J. and Rowell, J. (2003) *Social Vulnerability, Sustainable Livelihoods and Disasters.* London: DFID.
Carpenter, S.R., Westley, F. and Turner, M.G. (2005) Surrogates for resilience of social-ecological systems. *Ecosystems* 8 (8), 941–944.
Coleman, J. (1988) Social capital in the creation of human capital. *American Journal of Sociology* 94, 95–120.
Cutter, S.L. (1996) Vulnerability to environmental hazards. *Progress in Human Geography* 20 (4), 529–539.
Cutter, S.L. and Emrich, T. (2006) Moral hazard, social catastrophe: The changing face of vulnerability along the hurricane coasts. *The Annals of the American Academy of Political and Social Science* 604 (1), 102–112.
Cutter, S.L., Boruff, B.J. and Shirley, W.L. (2003) Social vulnerability and environmental hazards. *Social Science Quarterly* 84 (2), 242–261.
Cutter, S.L., Barnes, L., Berry, M., Burton, C., Evans, E., Tate, E. and Webb, J. (2008) A place-based model for understanding community resilience to natural disasters. *Global Environmental Change* 18 (4), 598–606.
Gallopin, G.C. (2006) Linkages between vulnerability, resilience and adaptive capacity. *Global Environmental Change* 16 (3), 293–303.

Gruppo di Lavoro MPS. (2004) *Redazione della mappa di pericolosità sismica prevista dall'Ordinanza PCM del 20 marzo 2003 n.3274 All. 1. Rapporto conclusivo per il Dipartimento della Protezione Civile, INGV, Milano-Roma* [online]. See http://zonesismiche.mi.ingv.it/.

Gunderson, L.H. (2000) Ecological resilience-in theory and application. *Annual Review of Ecology and Systematics* 31, 425–439.

Hall, C.M., Prayag, G. and Amore, A. (2017) *Tourism and Resilience: Individual, Organisational and Destination Perspectives*. Bristol: Channel View Publications.

Holling, C.S. (1973) Resilience and stability of ecological systems. *Annual Review of Ecology and Systematics* 4, 1–23.

Holling, C.S. (1996) Engineering resilience versus ecological resilience. In P. Schulze (ed.) *Engineering within Ecological Constraints* (pp. 31–44). Washington D.C.: National Academy Press.

Holling, C.S. (2001) Understanding the complexity of economic, ecological and social systems. *Ecosystems* 4 (5), 390–405.

Jansen, M.A., Schoon, M., Ke, W. and Borner, K. (2006) Scholarly networks on resilience, vulnerability and the human dimension of global environmental change. *Global Environmental Change* 16 (3), 240–252.

Keck, M. and Sakdapolark, P. (2013) What is social resilience? Lessons learned and ways forward. *Erdkunde* 67 (1), 5–19.

Lorenz, F. (2010) The diversity of resilience: Contributions from a social science prospective. *Natural Hazards* 67 (1), 7–24.

Magis, K. (2010) Community resilience: An indicator of social sustainability. *Society and Natural Resources* 23 (5), 401–416.

Norris, F.H., Stevens, S.P., Pfefferbaum, B., Wyche, K.F. and Pfefferbaum, R.L. (2008) Community resilience as a metaphor, theory, set of capacities, and strategy for disaster readiness. *American Journal of Community Psychology* 41 (1–2), 127–150.

Orchiston, C. (2010) Tourism and Seismic Risk: Perceptions, Preparedness and Resilience in the Zone of the Alpine Fault, Southern Alps, New Zealand. PhD thesis, University of Otago.

Orchiston, C. (2012) Seismic risk scenario planning and sustainable tourism management: Christchurch and the Alpine Fault zone, South Island, New Zealand. *Journal of Sustainable Tourism* 20 (1), 59–79.

Orchiston, C. (2013) Tourism business preparedness, resilience and disaster planning in a region of high seismic risk: The case of the Southern Alps, New Zealand. *Current Issues in Tourism* 16 (5), 477–494.

Pendall, R., Foster, K.A. and Cowell, M. (2010) Resilience and regions: Building understanding of the metaphor. *Cambridge Journal of Regions, Economy and Society* 3 (1), 71–84.

Putnam, R.D. (1993) *Making Democracy Work: Civic Traditions in Modern Italy*. Princeton, N.J.: Princeton University Press.

United Nations Development Programme (UNDP) (2004) *A Global Report. Reducing Disaster Risk: A Challenge for Development*. New York: UNDP.

United Nations International Strategy for Disaster Reduction (UNISDR) (2004) *Living With Risk. A Global Review of Disaster Reduction Initiatives*. Geneva, Switzerland: UNISDR.

Vogel, C. and O'Brein, K. (2004) Vulnerability and Global Environmental Change: Rhetoric and Reality. *GECHS Information Bulletin 'Aviso'* 13/2004.

Voss, M. (2008) The vulnerable can't speak. An integrative vulnerability approach to disaster and climate change research. *Behemoth-A Journal of Civilization* 1 (3), 39–56.

Walker, B.H. and Meyers, J.A. (2004) Thresholds in ecological and social ecological systems. A developing database. *Ecology and Society* 9 (2), 3.

Walker, B., Holling, C.S., Carpenter, S.R. and Kinzig, Z.A. (2004) Resilience, adaptability and transformability in social-ecological systems. *Ecology and Society* 9 (2), 5.

Williams, F. and Drury, S. (2009) Psychosocial resilience and its influence on managing mass emergencies and disasters. *Psychiatry* 8 (8), 293–296.

3 Crisis Communication Systems and Earthquake Preparedness of Tourism Sectors in LDCs: A Study on Nepal

Subhajit Das and Premangshu Chakrabarty

Introduction

Disasters have become an integral fabric of the social system. However, not all places in the world face similar types of disasters and their distinctive consequences. Some of them can be forecast with high precision prior to the occurrence and some not. Disasters such as earthquakes, where early forecasting has always been quite uncertain, are more prone to cause huge loss of lives and resources. The damage in such cases can only be minimised but cannot be stopped. As a result, effective disaster preparedness is vital (Drabek, 2007; Brannigan, 2015). It is more crucial than ever to resolve how to get prepared for responding to the sudden incidents like earthquake in order to cope with the disastrous consequences, and to help ensure that a resilient recovery is possible. Preparedness, response and recovery are inter-connected in the framework of any effective disaster management system as they signify a sequential course of action during the time of crisis. However, coming from outside of a destination and being unfamiliar with the place of crisis, tourists are always susceptible to any sudden calamities like earthquakes (Murphy & Bayley, 1989; Burby & Wagner, 1996; Kunwar & Limbu, 2015). Furthermore, earthquakes are always considered to be very critical in tourism sector disaster response because of damage to the touristscape, potential tourist casualties and injuries and the subsequent impacts on destination image (Ritchie, 2004; Santana, 2004; Glaesser, 2006).

The effectiveness of a crisis communication system (CCS) in association with crowd sourcing (CS) and social sensing (SS) to minimise the disaster induced crisis effects in tourism and other sectors has become increasingly significant (Ritchie, 2004; Glaesser, 2006; Laws *et al.*, 2006; Mansfeld, 2006; Ali *et al.*, 2011; Estellés-Arolas & González-Ladrón-De-Guevara, 2012; Ritchie & Campiranon, 2014; Liu *et al.*, 2015; Wang *et al.*, 2015; Mair *et al.*, 2016). The United Nations World Tourism Organisation (UNWTO) also recommends that earthquake-prone Less Developed Countries (LDCs) incorporate CCS in their disaster management plans and policies to enhance the disaster preparedness of stakeholders at tourist destinations (UNWTO, 2011; Association of Southeast Asian Nations (ASEAN), 2015). However, no specific research has yet been recorded that addresses the issues and challenges pertaining to the development of such technological systems in the LDCs to manage earthquake events. This chapter aims to focus on the major hindrances and potentials for LDCs in implementing CCS (in association with CS and SS) in the tourism sector, particularly during earthquake incidents. The chapter begins with a discussion on the earthquake-tourism interplay in the context of tourism crises, followed by the concepts of crisis communication system, crowd sourcing and social sensing and their role in disaster management plans. After that, some critical issues are raised in relation to adopt such strategies in the disaster management plans for the tourism in the LDCs. The chapter concludes with the Nepal case study to provide empirical insight into the potential prospects of LDCs adopting CCS during earthquake incidents.

The Earthquake-Crisis-Tourism Interface

Tourist destinations are always exposed to the potential occurrence of disasters of various types. Although there is a continuous discourse on defining disaster and crisis, a comprehensive understanding of disaster in the context of tourism business is yet to be developed (Glaesser, 2006). Although the frequency of natural disasters, such as earthquakes, has not increased at the global scale, people have become more exposed to such events because of population growth, rapid urbanisation, unplanned development in hazard-prone areas and ever-increasing global economic pressures (Faulkner, 2001). Disasters affect tourist destinations negatively at different levels of tourism system ranging from the organisational stakeholders to the destination image itself (Santana, 2004; Glaesser, 2006). However, the effects could be minimised if the event could be forecast in advance and there would be sufficient time to escape from the disastrous consequences. Sudden natural events, such as an earthquake, are considered as special types of disasters, because neither is early prediction certain nor immediate escape possible. Such disasters are sometimes described as no-escape natural disasters (NEND) (Huan *et al.*, 2004). Being a NEND, earthquakes causes huge damage to any tourist destination (Schneid & Collins, 2000; Mazzocchi & Montini, 2001;

Huang & Min, 2002; Avvenuti *et al.*, 2014), especially when no effective preparedness planning framework is available to cope with such crisis situation.

Disasters like earthquakes, induce sudden and random crisis situations at the tourist destination where the tourism stakeholders and enterprises have very little control over it and tend to respond at the post-disaster stage, either with the 'contingency plans already in place or through reactive response' (Scott & Laws, 2006: 152). On the other hand, a crisis situation may also arise in the post-disaster phase due to the failure of 'management structure and practices' (Scott & Laws, 2006: 152) to adopt an effective plan to change the disruption even after knowing that anytime the normal operations of tourism business may be hampered by a disaster (Faulkner, 2001; Sönmez *et al.*, 1994).

A tourism crisis refers to any situation that affects and hampers the existing normal tourism business operations due to the negative image of tourists with respect to the concerned destination in regards to safety, attractiveness and other issues; which results in a downturn in the local tourism economy and an uncertain future because of a sustained decline in tourist arrivals and associated expenditures. In a tourism context, crisis and disaster are often epitomised as chaos phenomena (Ritchie, 2004), as tourism business always waits for a disaster induced crisis situation in the pre-crisis phase or 'prodromal modea' (Fink, 1986: 15). Sometimes, earthquakes may provide a positive scope to tourism managers and stakeholders to realign and re-organise the touritscape and other tourism operations in a more attractive way keeping in mind the new market demand for the potential tourism products. Therefore, tourism crisis and earthquake may have a transformational connotation in terms of their positive and negative outcomes (Faulkner, 2001). However, the effective management of tourism crisis can ensure more resilient tourism operations by reducing the risk and uncertainty of an earthquake event.

CCS for Earthquake-Prone Destination Management

Crisis communication is referred as any 'verbal or non-verbal' communication and response that is meant to share the crisis information at the intra- or inter-organisational level (Coombs, 1999). However, awareness of the importance of crisis communication in relation to tourism management appears to be a very recent phenomenon. According to Glaesser (2006), negative event induced crisis (like natural catastrophes, earthquake, cyclone) should be handled separately from a crisis caused by human activities. A proactive approach is more common in human induced crisis management, as the levels of psychological and managerial preparedness are subject to the anticipation of potential long term effects of possible future negative events. Notwithstanding, a reactive approach is potentially more applicable to natural crisis management, as the crisis period starts immediately after the occurrence and detection of the negative events. This book chapter considers tourism crisis as a consequence of disasters in general and earthquakes in particular.

During the onset of an earthquake sequence, it is essential to institutionalise a reactive management approach in order to reduce the severity of tourism crisis. Based on the work of Faulkner (2001), Ritchie (2004) proposed a tourism crisis management framework in which crisis communication is considered as an indispensable tool for strategic control over the natural disaster induced tourism crisis. When a tourism crisis has been caused as the result of an earthquake, the 'game' of blame and responsibility becomes very crucial (Harrison, 2016). An effective destination management framework at an earthquake prone destination should therefore involve a disaster prevention plan first, followed by the crisis communication strategy at the response stage. The success of the crisis communication strategy lies in the positive and cooperative interrelationships among the tourism and other stakeholders who are actively participating in channelling the information on earthquake impacts to the right management authority at the response phase (Fishman, 1999).

The UNWTO and the ASEAN have come up with the guiding manuals for the development of effective crisis communication systems for earthquake prone tourist destinations (UNWTO, 2011; ASEAN, 2015). CCS should be developed as an integrated system of two specific teams, one is the Crisis Management Team (CMT) and the other is the Crisis Communication Team (CCT). The CMT should be the central or nodal tourism management authority that will take care of the overall management of tourism crisis situations, whereas the CCT would be the dedicated sub-teams of CMT to look after the flow and communication of crisis information to the CMT from the tourism stakeholder groups. To make this system of crisis communication effective, CMTs are recommended to have a pre-planned crisis communication policy that will train the members of CCTs and assign pre-defined responsibilities upon them, so that no chaotic situation may arise during the outbreak of any earthquake event.

CCS depends on three forms of communication: active, passive and exponential. Active and passive forms are the most crucial during the time of immediate response actions in a crisis situation. Active or push communication occurs when the victims or responders or the concerned tourism stakeholders communicate the crisis situation to other individuals who are more or less directly known to them. In passive or pull communication, a dedicated website is designed by the tourism management or other authorities to call for the submission of information about crisis situation from the 'intended audience'. Exponential communication involves the gradual passing of information to multiple recipients from a specific source. It is considered to be the fastest form of communication which can immediately make many people aware of the crisis situation within a very short span of time. However, fake information may cause a chaotic situation among tourists (UNWTO, 2011). For all such communication media, the 'social web' is regarded as having a key role in CCS because of its potential capacity to effectively transfer crisis information to many recipients in a very short time period as compared to traditional media such as television, newspapers, and radio. In addition, the traditional media channels also rely on social media and other networking platforms, such as Facebook, Twitter and Instagram, to receive information from active and passive communication sources and then broadcast crisis information. Such developments are also related to the growth of Crowd Sourcing and Social sensing activities (Besaleva & Weaver, 2013).

Crowd sourcing

Crowd sourcing has been defined in many ways. In seeking to integrate the various conceptualisations of crowdsourcing Estellés-Arolas and González-Ladrón-De-Guevara (2012) suggested that the notion of crowd could refer to any group of individuals, either experts or non-experts, who respond to the open call for a particular crisis management initiative to solve a problem, sometimes in exchange of monetary incentives or social recognition or the entertainment values. Sourcing refers to the strategic practices which aim to involve people through the internet in order to receive authentic information about the problem for which the open call has been made.

During an earthquake (or any other no-escape disaster) at a tourist destination, the open call may be initiated first by the CMT. In such open calls, people may be requested to send authentic information about the earthquake impacts on tourists and tourism infrastructures. Then the CCT, along with other group of stakeholders (both tourists and the locals), may directly respond to such calls with the help of social media and other social networking platforms such as Facebook and Twitter. As the first few hours are very crucial in the case of no-escape disaster events, crowd sourcing fits into the framework of CCS as it communicates the relevant crisis information about earthquake induced damages and losses to the CMTs for immediate rescue actions and the emergency resource mobilisation to the affected sites.

Social sensing

Social sensing considers humans as sensors that surpasses the limitations of any traditional physical sensor (Rindfuss & Stern, 1998). Individuals play many roles in social sensing systems (Aggarwal & Abdelzaher, 2013). They may act as real-time sources of information, or they may even operate as physical sensors by providing pictures and geo-tagged information of a disaster event. The pervasiveness of social sensing is attributed to three major contemporary societal trends. Firstly, the commercial proliferation of sensing devices (smart phones, GPS enabled tracking devices and other social sensing applications) that are being accessed by many people around the world. Secondly, ubiquitous access to mobile internet connectivity. Thirdly, the mass popularity of social media applications and information dissemination channels e.g. Facebook, Twitter, YouTube (Wang et al., 2015). Information, being shared through these social sensing platforms, needs to be systematically processed to extract the embedded 'whispering' of social perspectives. As a result, researchers are regularly coming up with new sophisticated computational and mathematical models to handle the huge volume of social sensing data (Wang et al., 2015).

Early research mostly witnessed the application of social sensing to monitoring and managing vehicular traffic system although they are now being applied to emerging fields such as health care services, the rescue of hikers and providing safe exit to the tourists during earthquake (Wang et al., 2015). When a tourist destination is affected by an earthquake and its subsequent aftershocks, all the emergency services and administrative bodies are immediately mobilised for damage control and rescue. The effectiveness of rescue operations lies in the authentic information regarding the situations of earthquake stricken locations where rescue becomes the first priority. The systematic study of individuals' response may help destination managers not only to mobilise the rescue team at the right place at right time, but also to gaze whether the taken actions are beneficial to the victims or not. The progress of relief activities could also be monitored by analysing how people are reacting to the emergency operations. Nevertheless, the application of social sensing also brings in many challenging issues like privacy of respondents, authenticity of shared information, coordinate accuracy, and the 'back-end challenges' while predicting and modelling the available social sensing data (Rindfuss & Stern, 1998; Ali et al., 2011; Aggarwal & Abdelzaher, 2013; Liu et al., 2015; Wang et al., 2015).

Both crowd sourcing and social sensing are embedded in the broader framework of CCS. Crowd sourcing works as a passive mode of information transfer to the CMTs through active participation of trained, pre-decided CCTs and other concerned tourism stakeholders, and social sensing opens up the platform for active and exponential mode of crisis communication through the complex network of the 'social web'.

LDCs' Readiness to Adopt CCS for Destination Management

National culture plays a very crucial role in human behaviour in disasters (Mora, 2013). The undertaking of best disaster management practices during disasters can help reduce the risk of a tourism crisis by create and reinforcing an image of a safe and competent country (Smits

& Ezzat Ally, 2003). However, there is a fundamental divide between the developed and developing or Less Developed Countries (LDCs) in terms of their approaches to handle tourism crisis during disaster events like earthquakes. Developed countries are considered to be 'low context' societies, characterised by explicit forms of knowledge and short-term relationships. Their management strategies are more open to accept the external opinions and universal principles. Whereas, LDCs are often framed as 'high context' societies in terms of implicit form of knowledge, long-term intra-relationships within society and inter-relationship with the environment. Such components cast their impacts in being less open to the external influences and limited to the adoption of principles of tourism crisis management even in case of earthquake preparedness (Adler & Gundersen, 2007).

LDCs are lagged behind in adopting developed country crisis management frameworks due to many limiting factors. It is quite problematic to expect that the LDCs are necessarily ready to adopt the architecture of CCS for destination management (Schmidt & Berrell, 2007). The hindrances in this context could be attributed to many factors: uncontrolled encroachment of human settlement in environmentally fragile areas due to cheaper cost of land resources; poverty; illiteracy; lack of economic and human resources; non-resilient physical and technological infrastructure; inaccessible remote habitats; leadership; attitudes of communities towards disaster; lack of co-ordination among tourism stakeholders; lack of trained tour operators; limited expertise in handling crowd sourcing and social sensing data; less preparedness for future disasters; and many more. Due to such limitations, LDCs are mostly found to react to the earthquake events on an *ad hoc* basis without a comprehensive pre-planned strategy to cope with an emergency. Nonetheless, the mental barriers and mind-sets of different social groups in the LDCs makes the crisis situation more chaotic and unmanageable (Campiranon & Scott, 2007; Mair *et al.*, 2016). However, in spite of many limitations, LDCs are considered to be very rich in the culture of participatory development which could assist in the adoption of effective CCS during earthquake emergencies (Schmidt & Berrell, 2007; Marahatta, 2012). Keeping the participatory culture of LDCs in mind, ASEAN modifies the CCS guidelines of UNWTO on a regular basis to integrate the tourism management bodies with tourism stakeholder groups at the grass root level, so that communities can actively take part in the responsive framework of CCS (ASEAN, 2015).

Managing Tourism Sector Through CCS: The Nepal Experience

Nepal is one of the most poverty stricken countries in the world and has been on the United Nations' list of LDCs since 1971. The country is the home of the highest peak of world – Mount Everest – and attracts a considerable number of tourists throughout the year. However, Nepal is also infamous for the frequent occurrences of natural hazards such as flood, landslide and earthquake. There is a substantial literature detailing why Nepal is an earthquake-prone country and the frequency of earthquake events that might be expected in the future (Pandey et al., 1999; Bollinger et al., 2014; Ministry of Home Affairs, Disaster Preparedness Network-Nepal, 2015; Center for Excellence in Disaster Management & Humanitarian Assistance, 2015). With a high chance of earthquake events in the future, the most challenging task of the Nepal government is to secure its image as a safe destination in the global tourism market.

The case study is based on the verification of earthquake impacts on tourism resources at different tourist sites in and around the Kathmandu valley. High ranking government officials and secretaries, specifically related to disaster management and tourism operations, were interviewed regarding the policy principles of the Nepal government towards future earthquake related tourism crises. They were also asked to share their experiences about how they handled the crisis situation during the 2015 earthquake in Nepal. Apart from such management authorities, other tourism stakeholders like tourist guides, tour operators, media persons and hotel managers were also interviewed with respect to their experiences in the last earthquake event and what they learnt for future preparedness. Local people were also taken into consideration in the interview schedules, so that their perception about earthquake events and concern for tourism operations could be ascertained.

CCS during 2015 earthquake

An earthquake with a magnitude of 7.6 on the Richter scale struck Nepal on 25 April 2015, and its subsequent aftershocks not only caused thousands of deaths, but also severely damaged the Nepalese economy with its immediate and longer term impacts. The disaster took place when Nepal was becoming increasingly open to outsiders, businesspersons and pleasure seekers. The architectural heritage sites, particularly the Durbar Squares of Kathmandu, Patan and Bhaktapur were devastated. With the aid of the Republic of China, the United States and other developed countries along with the South Asian Association for Regional Cooperation (SAARC) countries, restoration projects have already been started following the principle of *status quo* restoration and reinforcement. In order to protect and preserve the original historical information to its maximum, original components have been retained as much as possible in restoration. At the initial phase, the fear of earthquake and the pictures of damage of the heritage sites affected the focus of international tourists on Nepal. However, the situation is gradually changing as the memories of the earthquakes fade and the Nepal Tourism Board (NTB) is successfully launching alternative tourism products to regain the attention of world tourism market. The Nepal Tourism Board (NTB) coordinates between the Ministry of Tourism and the Directorate of Tourism to rationalise the promotional efforts of a diversified tourism product base.

Responsibility for disaster management in Nepal lies under the Ministry of Home Affairs which incorporates tourism and various sectors in its Disaster Risk Reduction (DRR) program. The Nepal Police as an organisation is assigned with the technical responsibilities for crisis response. Under the leadership of Dr Rajib Subba, Deputy Inspector General (Communication) of Nepal Police, crowd sourcing was initiated just after the 2015 earthquake. Though CCS becomes more complex during a large scale crisis, during 2015 earthquake Nepal police handled the situation with a strategic crisis communication approach which was based on five Cs – concern, clarity, control, confidence and competence (Subba, 2015). Along with various types of social media communications, text messaging, Facebook and Twitter were used for crowd sourcing.

Just after the earthquake, the first response came from Communication Directorate of Nepal Police requesting everybody to impart text messages on causalities, which was subsequently placed on the Facebook and Twitter accounts to receive information about the ground reality of damage and loss of human lives and properties. Information on the situation was also conveyed to field rescuers and relief teams via radio. During the earthquake, the social media platform Twitter demonstrated its effectiveness as a gateway to public engagement in participatory crisis management (Subba & Bui, 2017). A project, named *Sankatmochan* (meaning the saver) which also served as the missing persons' website, was launched by the Nepal Police based on the information received from crowd sourcing.

Nepal is well known for its community radio network, this not only connects remote areas to so-called modern society, but also provides a voice to the rural and marginalised communities to place their opinions before the administration. Community radio has been functional in Nepal since 1997, with a considerable number of radio broadcasting licenses only being distributed to the private sector after 2007. As the name suggests, community radio is a non-profit radio broadcasting system which operates within a small territory or a village, and the ownership belongs to the concerned community itself that manages all of its broadcasting programmes. Based on its participatory approach, community radio effectively involves local people, mainly women, and broadcasts programmes for social benefits, mostly related to education, health, culture and the emergency situations (Myers, 2011). Mass access to internet and mobile phones has eventually strengthened the community radio network of Nepal in terms of involving local people and communicating the community diaspora even during the crisis situations (Myers, 2011). During 2015 earthquake community radio played a very important role in transferring casualty (both people and property) and safety information to the remote villagers and the central administration of Nepal. The Nepal Police, disaster management authorities and tourism department were in touch with the community radio channels so as to assess the severity of the 2015 earthquake in rural Nepal. Such community radio channels not only fed the administrative authorities with the live situation of damage supported by photo and videos, but also communicated the affected tourists' location and trekkers' situation. After great efforts of filtering the information of community radio channels, the Nepal Tourist Police were also able to rescue many tourists from the remote trekking trails and many village destinations. Although, Nepal lacked the infrastructure of social sensing to process the 'push communication' in social media like Facebook and Twitter, Community Radio worked as an alternative source of social sensing in Nepal during 2015 earthquake.

CCS in post-2015 Nepal

With the lesson of the 2015 earthquake, the Nepal Government is now giving more emphasis on developing an effective CCS, mainly based on crowd sourcing, as social sensing is still in its infancy. A number of international and local NGOs like NSET (Nepal Society for Earthquake Technology) and CORD (Centre of Resilient Development) are presently operating in Nepal to develop CCS awareness and community preparedness for future earthquake events. These NGOs are working closely with the government agencies to provide community training and support with respect to infrastructure (building codes of conducts) as well as psychological preparedness. It is noteworthy to mention that about 70% of their activities are presently part of the CCS, mostly in building the community response by developing human resources to cope with future earthquakes. For example, a dedicated television programme, called *Sankalpa,* is being broadcasted by Nepal Television on every Thursday for half an hour to train rural communities for future disaster response. Tourist guides are also being trained to carry rescue equipment and high-energy food while escorting tourists in natural areas. At cultural sites, guides are trained to regulate tourists' movement with warnings of post-2015 condition of the traditional houses and the potential threat of building collapse.

Nepal is still lacking an integrated framework of CCS in the tourism sector due to the cognisable coordination gap between the Tourism Ministry and the disaster management division of Ministry of Home Affairs. Although, the NTB maintains the liaison with tourism ministry, the development of a tourist-centric disaster management strategy framework in Nepal is still in its infancy stage. During field survey, it was noted that there is lack of disaster awareness and preparedness among the Nepal Tourist Police personnel, and they do not have any pre-defined model of action to assist tourists at the time of future earthquake events. They are more familiar with the collective mobilisation of different government departments of Nepal and the ad-hoc basis rescue and relief activities during natural disasters. Although, secretary level officers from tourism ministry are involved in the central disaster management cell of Nepal as a nodal officer, the integrated course of response to future earthquakes still appears unstructured and amorphous. Nepal is in dire need of a structured policy framework for the CCS which will integrate the tourism sector with central disaster management.

Conclusion

The functional structured of CCS in LDCs should always be designed differently to that of developed countries. It is better to accept the limitations of LDCs in developing CCS with modern technology. Alternative communication strategies, e.g. community radio, and cost effective data processing system should also be considered for the best information output to tackle the tourism crisis during future earthquake events. The development of a dedicated CCS for the tourism sector in LDCs is retarded by the scarcity of resources and human expertise. Therefore, instead of individualistic approach to develop a dedicated CCS for tourism sector, it is more rational to adopt a holistic approach in LDCs which will serve both the hosts and guests. However, dedicated response teams for the rescue of tourists are invariably required under the policy framework of LDCs to safeguard the image of a safe and secure destination. Moreover, mass level involvement of local people may also hamper the success of CCS in LDCs, as they often lack technical proficiency and response expertise and need to be trained and made prepared for the CCS and earthquake events by NGOs in collaboration with government departments.

Acknowledgements

We pay our sincere gratitude and thanks to Mr H.E. Eaknarayan Aryal, the Consul General of Nepal in Kolkata, Mr K.B. Raut, Joint Secretary, Ministry of Home Affairs, Mr K.B. Shah of NTB and Dr Rajib Subba of Nepal Police. Without their help, this work would not have been completed. We are also thankful to all of the Government officials, members of NGOs and other respondents who cordially cooperated with us during our field visit in Nepal. Finally, Dr Das sincerely acknowledges the contribution of Presidency University for providing sufficient funds for field work under the FRPDF scheme.

References

Adler, N.J. and Gundersen, A. (2007) *International Dimensions of Organizational Behavior*. Sydney: Thomson South-Western.

Aggarwal, C.C. and Abdelzaher, T. (2013) Social sensing. In C.C. Aggarwal (ed.) *Managing and Mining Sensor Data* (pp. 237–297). New York: Springer US.

Ali, R., Solis, C., Salehie, M., Omoronyia, I., Nuseibeh, B. and Maalej, W. (2011) Social sensing: When users become monitors. In *Proceedings of the 19th ACM SIGSOFT Symposium and the 13th European Conference on Foundations of Software Engineering* (pp. 476–479). ACM.

ASEAN. (2015) *ASEAN Tourism Crisis Communications Manual: Incorporating Best Practices of PATA and UNWTO*. Jakarta, Indonesia: ASEAN.

Avvenuti, M., Cresci, S., Marchetti, A., Meletti, C. and Tesconi, M. (2014) EARS (earthquake alert and report system): A real time decision support system for earthquake crisis management. In *Proceedings of the 20th ACM SIGKDD International Conference on Knowledge Discovery and Data Mining* (pp. 1749–1758). ACM.

Besaleva, L.I. and Weaver, A.C. (2013) Applications of social networks and crowdsourcing for disaster management improvement. In *2013 International Conference on Social Computing*, IEEE (pp. 213–219).

Bollinger, L., Sapkota, S.N., Tapponnier, P., Klinger, Y., Rizza, M., Van Der Woerd, J., Tiwari, D., Pandey, R., Bitri, A. and Bes de Berc, S. (2014) Estimating the return times of great Himalayan earthquakes in eastern Nepal: Evidence from the Patu and Bardibas strands of the Main Frontal Thrust. *Journal of Geophysical Research: Solid Earth* 119 (9), 7123–7163.

Brannigan, M.C. (2015) *Japan's March 2011 Disaster and Moral Grit: Our Inescapable In-between*. Lanham: Lexington Books.

Burby, R.J. and Wagner, F. (1996) Protecting tourists from death and injury in coastal storms. *Disasters* 20 (1), 49–60.

Campiranon, K. and Scott, N. (2007) Factors influencing crisis management in tourism destinations. In E. Laws, B. Prideaux and K. Chon (eds) *Crisis Management in Tourism* (pp. 142–156). Wallingford: CABI.

Center for Excellence in Disaster Management and Humanitarian Assistance (2015) *Nepal Disaster Management Reference Handbook 2015* [pdf] See https://reliefweb.int/sites/reliefweb.int/files/resources/disaster-mgmt-ref-hdbk-2012-nepal.pdf.

Coombs, T. (1999) *Ongoing Crisis Communication – Planning, Managing and Responding*. London: Sage.

Drabek, T.E. (2007) Sociology, disasters and emergency management: History, contributions, and future agenda. In D.A. McEntire (ed.) *Disciplines, Disasters and Emergency Management: The Convergence and Divergence of Concepts, Issues and Trends in the Research Literature* (pp. 61–74). Springfield: Charles C. Thomas.

Estellés-Arolas, E. and González-Ladrón-De-Guevara, F. (2012) Towards an integrated crowdsourcing definition. *Journal of Information Science* 38 (2), 189–200.

Faulkner, B. (2001) Towards a framework for tourism disaster management. *Tourism Management* 22 (2), 135–147.

Fink, S. (1986) *Crisis Management: Planning for the Inevitable*. New York: American Management Association.

Fishman, D.A. (1999) ValuJet Flight 592: Crisis communication theory blended and extended. *Communication Quarterly* 47 (4), 345–375.

Glaesser, D. (2006) *Crisis Management in the Tourism Industry*. Amsterdam: Butterworth-Heinemann.

Harrison, S. (2016) *Disasters and the Media: Managing Crisis Communications*. London: Palgrave Macmillan.

Huan, T.-C., Beaman, J. and Shelby, L. (2004) No-escape natural disaster: Mitigating impacts on tourism. *Annals of Tourism Research* 31 (2), 255–273.

Huang, J.-H. and Min, J.C. (2002) Earthquake devastation and recovery in tourism: The Taiwan case. *Tourism Management* 23 (2), 145–154.

Kunwar, R.R. and Limbu, B. (2015) Tourism and earthquake: A case study of Nepal and Turkey. *Building Better Tourism with Renewed Strength, XXth NATTA Convention* 16–31.

Laws, E., Prideaux, B. and Chon, K.S. (eds) (2006) *Crisis Management in Tourism*. Wallingford: CABI.

Liu, Y., Liu, X., Gao, S., Gong, L., Kang, C., Zhi, Y., Chi, G. and Shi, L. (2015) Social sensing: A new approach to understanding our socioeconomic environments. *Annals of the Association of American Geographers* 105 (3), 512–530.

Mair, J., Ritchie, B.W. and Walters, G. (2016) Towards a research agenda for post-disaster and post-crisis recovery strategies for tourist destinations: A narrative review. *Current Issues in Tourism* 19 (1), 1–26.

Mansfeld, Y. (2006) The role of security information in tourism crisis management: The missing link. In Y. Mansfeld and A. Pizam (eds) *Tourism, Security & Safety: From Theory to Practice* (pp. 271–290). Amsterdam: Elsevier Butterworth-Heinemann.

Marahatta, P.S. (2012) Community-based approach to reduce earthquake vulnerability in Kathmandu Valley. *Indonesian Journal of Geography* 44 (2), 161–172.

Mazzocchi, M. and Montini, A. (2001) Earthquake effects on tourism in central Italy. *Annals of Tourism Research* 28 (4), 1031–1046.

Ministry of Home Affairs, Disaster Preparedness Network-Nepal. (2015) *Nepal Disaster Report 2015*. Kathmandu: Ministry of Home Affairs, Disaster Preparedness Network-Nepal.

Mora, C. (2013) Cultures and organizations: Software of the mind intercultural cooperation and its importance for survival. *Journal of Media Research* 6 (1), 65.

Murphy, P.E. and Bayley, R. (1989) Tourism and disaster planning. *Geographical Review* 79 (1), 36–46.

Myers, M. (2011) *Voices from Villages: Community Radio in the Developing World: A Report to the Center for International Media Assistance*. Washington DC: Center for International Media Assistance (CIMA).

Pandey, M., Tandukar, R., Avouac, J., Vergne, J. and Heritier, T. (1999) Seismotectonics of the Nepal Himalaya from a local seismic network. *Journal of Asian Earth Sciences* 17 (5–6), 703–712.

Rindfuss, R.R. and Stern, P.C. (1998) Linking remote sensing and social science: The need and the challenges. In National Research Council, Division of Behavioral and Social Sciences and Education, Board on Environmental Change and Society, Committee on the Human Dimensions of Global Change (eds.) *People and Pixels: Linking Remote Sensing and Social Science* (pp. 1–27). Washington, DC: National Academies Press.

Ritchie, B.W. (2004) Chaos, crises and disasters: A strategic approach to crisis management in the tourism industry. *Tourism Management* 25 (6), 669–683.

Ritchie, B.W. and Campiranon, K. (eds) (2014) *Tourism Crisis and Disaster Management in the Asia-Pacific*. Wallingford: CABI.

Santana, G. (2004) Crisis management and tourism: Beyond the rhetoric. *Journal of Travel & Tourism Marketing* 15 (4), 299–321.

Schmidt, P. and Berrell, M. (2007) Western and eastern approaches to crisis management for global tourism: Some differences. In E. Laws, B. Prideaux and K. Chon (eds) *Crisis Management in Tourism* (pp. 66–80). Wallingford: CABI.

Schneid, T.D. and Collins, L.R. (2000) *Disaster Management and Preparedness*. London: Lewis Publishers.

Scott, N. and Laws, E. (2006) Tourism crises and disasters: Enhancing understanding of system effects. *Journal of Travel & Tourism Marketing* 19 (2–3), 149–158.

Smits, S.J. and Ezzat Ally, N. (2003) 'Thinking the unthinkable' – Leadership's role in creating behavioral readiness for crisis management. *Competitiveness Review: An International Business Journal* 13 (1), 1–23.

Sönmez, S.F., Backman, S.J. and Allen, L. (1994) *Managing Tourism Crises: A Guidebook*. Clemson: Department of Parks, Recreation and Tourism Management, Clemson University.

Subba, R. (2015) Earthquake rescue: Twitter for crisis communication. *Republica*, 12 July. See http://admin.myrepublica.com/opinion/story/24454/twitter-effects.html.

Subba, R. and Bui, T. (2017) Online convergence behavior, social media communications and crisis response: An empirical study of the 2015 Nepal earthquake police twitter project. *Proceedings of the 50th Hawaii International Conference on System Sciences*, Hawaii.

Wang, D., Abdelzaher, T. and Kaplan, L. (2015) *Social Sensing: Building Reliable Systems on Unreliable Data*. Amsterdam: Morgan Kaufmann.

World Tourism Organization (UNWTO) (2011) *Toolbox for Crisis Communications in Tourism: Checklists and Best Practices*. Madrid: UNWTO.

4 Mitigating Earthquake and Tsunami Risks in Coastal Tourism Sites in Bali

I Nengah Subadra

Introduction

Earthquakes are one of the most common disasters in Indonesia since its archipelagos are located in an area of high seismicity. Almost all regions in Indonesia are susceptible to earthquake and tsunami disasters that have caused massive casualties (Nguyen *et al.*, 2015; Badan Meteorologi Klimatologi dan Geofisika [Meteorology, Climatology and Geophysics Agency], 2017). For example, the Aceh earthquake of 26 December 2004 reached a magnitude of 9.3 on the Richter scale and was followed by a tsunami killing at least 283,000 people and damaging thousand houses, buildings and public facilities along Aceh's coastal zones (Gioncu & Mazzolani, 2011). Such natural disasters serve as a reference to plan and develop integrated disaster mitigation systems that help enable the minimization of such losses.

Bali, in particular, has a very long history of earthquake and tsunami disasters. Nineteenth century earthquakes damaged infrastructure and led to the loss of thousands of lives. More recently, the northern and eastern regions of the island were hit by earthquakes and tsunamis in 1976 and 1979 resulting in many fatalities (Khomarudin *et al.*, 2010; Strunz *et al.*, 2011; Suppasri *et al.*, 2012). The high threat of Bali from earthquake disasters is due to the southern region of Bali directly facing the Indo-Australian plate and the northern region being located on the edge of fault rise of the ocean base. When such disasters occur, they can have major impacts on tourism development (Huang & Min, 2002). Preparedness for natural disaster mitigation is therefore essential and needs to be well planned in order to reinforce resilience and be able to deal with disasters, achieve appropriate response and actions when it occurs and assist in recovery (McCool, 2012). As a major international destination, it is therefore essential for Bali to have a particular agency which is able to coordinate and manage all tourism stakeholders such as government, hotel managements, local people and tourists in mitigating the risks of earthquake and tsunami disasters that threaten tourism sites. This is especially important for the southern coastal regions of Bali where most holiday resorts are located and the greatest number of tourists stay.

Bali is a small island in the Indonesian archipelago lying approximately eight degrees south of the equator with a total area of 5780 km^2 and inhabited by a population of close to four million people. Unlike the majority of Indonesia, Bali's population is over eighty percent Hindu. This situation makes Bali culturally distinct from neighbouring islands such as Lombok and Java which have Muslim majority populations; providing a cultural element that compliments other tourist motivations to visit the island (Subadra, 2015).

Bali is a major international holiday destination visited by millions of domestic and international tourists every year. Balinese culture is now central to tourism development in the island under provincial law by which Bali cultural tourism is to be used as the principal guidance in developing tourism in any regions of the island (Regional Regulation of Bali Province Number 2 of 2012). Subadra (2015) argues that tourism in Bali today is not all about culture, but has also been extended to natural based tourism, as witnessed in the rapid development of dive centres, rafting companies, interisland cruises and beach tourism together with new attractions, such as water boom, bungy jumping and all-terrain vehicle (ATV) rides.

Bali has been a major component of tourism to Indonesia for many years (Table 4.1). In the space of 50 years the number of foreign visitors to Bali has grown from 11,278 to over 6 million. In 2018 Bali accounted for almost 40% of foreign visitors to Indonesia and the number of visitors to Bali is now approaching 150% that of the number of permanent residents

Dating back to the initial stage of tourism on Bali island when the first international tourists visited Bali in the 1920s (Picard, 1996), tourism developments have been concentrated in the southern regions of Bali including Nusa Dua, Kuta and the Sanur area. These were also the locations in which the rapid tourism development occurred during Soeharto's New Order regime between 1966 and 1998 with infrastructural developments such as roads, international airport, harbours and hotel accommodation (Subadra, 2015). This period is also marked by the active promotion of the island by foreign authors who enthusiastically promoted the uniqueness of Balinese cultures and the island's beauty through creative branding as a 'last paradise' (Powell, 1982) and the 'island of a thousand temples' (Moore, 1970). The legacy of such branding remains to the present-day in terms of industry promotion and even the attraction reviews of tourists on Trip Advisor (Withnall, 2016).

As of 2018 the southern coastal region of Bali accounts for over 95% of classified hotel room accommodation on Bali and approximately 63%

Table 4.1 Number of foreign visitors to Indonesia and Bali, 1969–2018

Year	Indonesia Total	Growth (%)	Bali Total	Growth (%)
1969	86,067	–	11.278	–
1970	129,319	50.25	24.340	115.82
1971	178.781	38.25	34.313	40.97
1972	221.195	23.72	47.004	36.99
1973	270.303	22.20	53.803	14.46
1974	313.452	15.96	57.456	6.79
1975	366.293	16.86	75.790	31.91
1976	401.237	9.54	115.220	52.03
1977	456.718	13.83	119.095	3.36
1978	468.614	2.60	133.225	11.86
1979	501.430	7.00	120.139	−9.82
1980	561.178	11.92	139.695	16.28
1981	600.151	6.94	153.030	9.55
1982	592.046	−1.35	150.673	−1.54
1983	638.855	7.91	166.575	10.55
1984	700.910	9.71	188.833	13.36
1985	749.351	6.91	211.222	11.86
1986	825.035	10.10	243.354	15.21
1987	1.060.547	28.55	309.292	27.10
1988	1.301.049	22.68	360.413	16.53
1989	1.625.965	24.97	436.358	21.07
1990	2.051.686	26.18	489.710	12.23
1991	2.569.870	25.26	554.975	13.33
1992	3.060.197	19.08	735.777	32.58
1993	3.403.138	11.21	884.206	20.17
1994	4.006.312	17.72	1.030.944	16.60
1995	4.310.504	7.59	1.014.085	−1.64
1996	5.034.472	16.80	1.138.895	12.31
1997	5.184.486	2.98	1.230.316	8.03
1998	4.606.416	−11.15	1.187.153	−3.51
1999	4.600.000	−0.14	1.355.799	14.21
2000	5.064.217	10.09	1.412.839	4.21
2001	5.153.620	1.77	1.356.774	−3.97
2002	5.033.400	−2.33	1.285.842	−5.23
2003	4.467.021	−11.25	993.185	−22.76
2004	5.321.165	19.12	1472.190	48.23
2005	5.002.101	−6.00	1.388.984	−5.65

(Continued on next page)

Table 4.1 (Continued)

Year	Indonesia Total	Growth (%)	Bali Total	Growth (%)
2006	4.871.351	−2.61	1.262.537	−9.10
2007	5.505.759	13.02	1.668.531	32.16
2008	6.234.497	13.24	2.085.084	24.97
2009	6.323.730	1.43	2.385.122	14.39
2010	7.002.944	10.74	2.576.142	8.01
2011	7.649.731	9.24	2.826.709	9.73
2012	8.044.462	5.16	2.949.332	4.34
2013	8.802.129	9.42	3.278.598	11.16
2014	9.435.411	7.19	3.766.638	14.89
2015	10.406.291	10.29	4.001.835	6.24
2016	11.519.275	10.70	4.927.937	23.14
2017	14.039.799	21.88	5.697.739	15.62
2018	15.806.191	12.58	6.070.473	6.54

Source: Badan Pusat Statistik Provinsi Bali (Statistics of Bali Province) (2019a).

of non-classified hotel rooms and rooms in other accommodation. Growth in tourist visitation has also been accompanied by a massive increase in accommodation, provision of which has almost tripled between 2010 and 2018 (Table 4.2). Not surprisingly, the growth of tourism in the southern regions of Bali plays a great role in its regional development. In the Badung Regency (a term used to describe a second-level administrative division, directly administrated under a province, sometimes equivalent to a municipality) tourism serves as the main regional revenue generator which has meant that this region is the richest regency in Bali although negative impacts of tourism include traffic congestion, air pollution and the conversion of fertile cultivated land for tourism related development (Subadra, 2015). Significantly for the purposes of the present chapter, the tourism amenities and attractions are mostly located and developed in the southern coastal regions of the island which is also most at risk of earthquake and tsunami (Kelman et al., 2008). However, tourists experience more difficulties than local people with respect to security and disaster preparedness since, as new temporary visitors to the region, tourists have little knowledge of the region and are unfamiliar with the environments which will affect their response to an emergency (Garg, 2015; Jensen & Svendsen, 2016). Therefore, effective disaster mitigation planning which involves all tourism stakeholders, including tourists, is crucial in responding to disaster.

Another important reason for disaster mitigation in southern Bali is the location of core infrastructure, such as the international airport and the power station – the latter supplying all electricity generation for

Table 4.2 Number of rooms in classified hotels, non-classified hotels and other accommodation in the southern coastal region of Bali

Regions	2010	2012	2014	2018
Badung regency				
Classified hotels	16,027	18.613	23,172	44,571
Nonclassified hotels and other accommodation	7929	9112	10,168	28,371
Denpasar city				
Classified hotels	3415	3619	3480	5437
Nonclassified hotels and other accommodation	5034	5526	5451	7316
Total	32,405	36.870	42,271	85,695

Source: Based on hotels survey and number of rooms available in non-classified hotels and other accommodation by rooms in Bali; Badan Pusat Statistik Provinsi Bali (Statistics of Bali Province) (2019b, 2019c), no figures available for 2016.

the island, and both located in the high risk 'red zone' of the southern coastal region. The operation of such infrastructure is clearly crucial to tourism operations on the island. The international airport of I Gusti Ngurah Rai for instance, located in Tuban Village, Badung Regency, currently serves as the main embarkation and disembarkation for domestic and international tourists in Bali. The closures of the airport on 29–30 November 2017 as a result of the Mount Agung eruption, for example, resulted in hundreds of flight cancelations and forced all airline operations to close for two days (Syahbana *et al.*, 2019).

In the case of the Mount Agung eruption thousands of tourists who had booked to fly on those dates were unable to fly out and were required to patiently wait and remain for the airport authority to reopen airports. In the emergency, some tourists were well treated by airline operators while others were unattended and waited around the airport for their flights to be rescheduled. It was very unclear as to where responsibility for tourist welfare ought to lie in such a situation, i.e. Bali's provincial government, airline operators, hotels or travel agencies. There was no coordination among stakeholders which left many unattended and disappointed. Similar issues affected those tourists who could not land on the island and were diverted to the nearest international airports located in the neighbouring islands, i.e. Lombok International Airport in Lombok Island and Juanda International Airport in Surabaya, Java. Given the travel chaos that occurred as a result of the eruption, it is therefore extremely important for Bali as an international tourist destination to develop a disaster mitigation system able to manage any kinds of natural disaster including volcanic and tectonic earthquakes and tsunami. Overall tourism stakeholders in Bali appeared to have little awareness of natural disaster mitigation issues as they failed to manage the outcomes of the Mount Agung eruption. As McCool (2012) argues, the less awareness hotels have of disaster mitigation planning, the lower is their own level of disaster preparedness and corresponding resilience.

Awareness of security and safety is not only essential for the tourists visiting the destination but also for hotel management, travel agencies and local people inhabiting the region.

Earthquake and Tsunami Risks on Bali

The location of Bali makes the island extremely vulnerable to earthquake and tsunami disasters. Earthquakes remain unpredictable and may arrive with no initial warning (Gioncu & Mazzolani, 2011). For example, on 14 July 1976 a magnitude earthquake of 6.5, known locally as *Gempa Seririt*, hit northern coastal Bali and destroyed the entire subdistrict of Seririt and severely affected other neighbouring districts and caused thousands of people to lose their properties (Utomo, 2011). Historically, Bali has suffered such disasters many times and therefore the preparedness of the local people and government in mitigating earthquake and tsunami risk is essential. Over 350,000 permanent residents in Bali are potentially at risk from a tsunami, of which 65% reside on the Southern coast and 35% in the northern coastal regions such as Singaraja, Lovina, Seririt and Celukan Bawang (Bali Badan Penanggulangan Bencana Daerah (BPBD), 2017a). This number excludes the number of tourists staying in the hotels, villas and other types of accommodation located in the two coastal zones or touring those regions.

Based on the histories of earthquakes and tsunamis in Bali, the Regional Disaster Management Agency found that any tsunami wave usually reaches the Southern coastal regions of Bali within 20 to 30 minutes of an earthquake; and in the Northern coastal areas in less than five minutes. This suggests that people living along the coastal regions of the island are highly vulnerable and have only a very limited time to reach safety. Therefore, a disaster risk mitigation system is integral to coping with such vulnerability.

Nguyen *et al.* (2016) identified two types of disaster vulnerability: physical vulnerability, which covers structure, infrastructure and natural damages; and social vulnerability, that includes the reliance on tourism by local people and business and also tourists due to their lack of knowledge of potential local hazard risks in the visited site and their lack of understanding of the local language. Indeed, due to the presence of tourists, earthquake and tsunami disaster mitigation in tourism destinations like Bali can be more complex than those that occur in non-tourism areas.

Earthquake and tsunami risk mitigations: Policy and practice

Because of the social and economic threats they pose natural disasters such as earthquake and tsunami have become a priority project for Bali's government. Jensen and Svendsen (2016) argue that both local tourism stakeholders and tourists visiting the destination should have awareness of potential disasters threatening a destination. Such awareness corresponds to current central Indonesian government policy on disaster mitigation stipulated under Act of Republic of Indonesia Number 24 of 2007 (Undang-Undang Republik Indonesia Nomor 24 Tahun 2007) which legally requires the minimization of the impacts of disasters.

The Indonesian government implements this particular Act via a specific bureau, the National Disaster Management Agency (BNPB) (*Badan Nasional Penanggulangan Bencana*). This body is responsible for formulating policy on disaster mitigation; the management of people displaced as a result of disaster; and coordinating and implementing integrated and holistic disaster mitigation. However, the size of the Indonesian archipelago has meant that provincial and district Regional Disaster Management Agencies (BPBDs) (*Badan Penanggulangan Bencana Daerah*) have also been created at a regional level. 'The sheer size and scale of Indonesia, frequency of disasters and logistical remoteness of many areas means that the critical first hours and even days of responses remain heavily dependent on local and provincial monitoring and response capacity, making BPBDs essential in some form' (Hodgkin, 2016: 32).

In the case of Bali, the provincial government of Bali has a Regional Disaster Management Agency which is specifically responsible for mitigating disasters occurring within the province of Bali. This agency plays three vital roles in natural disaster mitigation in Bali including coordinator, evacuator and commander. As coordinator, this agency coordinates with other government authorities responsible for disaster mitigation including the Meteorology, Climatology and Geophysics Agency where official disaster informational notices are initially received and broadcast; the Department Social and Welfare that deals with displaced persons; the Department of Public Works which is responsible for infrastructure; the Department of Regional Revenue which manages and provides the provincial budget for disaster funding; and the police department and army which is responsible for order and security. Furthermore, the Regional Disaster Management Agency of Bali acts as the key evacuators in case of disaster with its officials backed up by the police and army in order to evacuate people from hazardous zones, preparing shelters for displaced persons and supplying displaced persons' basic needs. Lastly, as commander, the Bali BPBD is responsible for providing official disaster information and coordinating the actions of disaster mitigation stakeholders (BPBD, 2017a).

The Bali BPBD also controls nine regional disaster mitigation agencies that are based in each regency of the island. As such, when any disaster occurs in a regency in Bali then it will be immediately coordinated by the BPBD. These includes the collection and distribution of funds for victims and the provision of search and rescue teams where necessary. Although provincial in scope the Bali BPBD follows national policies and regulations concerning disaster mitigation operations. This is aimed at ensuring that disaster mitigation is well executed by the assigned agency to protect the local people from threats, risks and impacts of disasters. The operation of disaster mitigation includes pre-disaster management, immediate-disaster management and post-disaster management (BPBD, 2017a).

Pre-disaster Mitigation Programs

The Bali BPBD has initiated pre-disaster management programs that cover formation of a disaster crisis centre, earthquake and disaster mapping, installing evacuation zone signage, and disaster management training. These programs involve a number of stakeholders including government, local people, tourism industries, academics and tourists.

In the effort to maximize a quick response to disasters, Bali's BPBD has formed a particular 'Crisis Centre' unit which operates and monitors 24/7 and provides an early warning system and disaster management control. This unit collects and conveys information on any potential disasters that may occur to the appropriate government authorities and aids in reducing disaster risks by taking the immediate steps to minimize casualties (see also Glaesser, 2003).

In order to reduce the vulnerability of the southern coastal regions of Bali to tsunami related disaster, the BPBD of Bali has mapped the areas most likely affected by tsunami. In term of pre-disaster mitigation, these maps have been used to help government and others to better understand tsunami threats in terms of their location, and the affected population, public services and territory. Additionally, the agency has also posted a tsunami warning board and determined evacuation zones and assembly points completed with signage on the coastal roads in Denpasar City and Badung Regency. These efforts are primarily aimed at preparing appropriate directions for the local people and tourists in a tsunami emergency (BPBD, 2017a).

In addition to government funded signage, government policy and regulations also require hotels sited along the coast to display evacuation information. The instalment of signage is usually a part of a hotel safety program that aims to show the sense of awareness of the hotel management on the danger of tsunami and the significance of a mitigation program to staff and tourists staying in the hotel. Prama Sanur Beach Hotel for instance, displays at least five tsunami disaster signs along the beach front located off Sanur Beach (Plate 4.1). The installed signage shows clear directions for guests and staff regarding where to go to and assemble.

The Bali BPBD also cooperates with another government body, the Geophysics, Climatology, and Meteorology Agency, to develop tsunami

Plate 4.1 Hotel displaying tsunami evacuation signage at Sanur Beach, 2017
Source: Author.

early warning systems. As of early 2018 there were nine tsunami early warning system devices installed with the number being increased each year depending on budget availability to eventually cover at least in nine major points around the island including Sanur, Serangan (Denpasar City), Tanjung Benoa, Nusa Dua, Kedonganan, Kuta, Seminyak (Badung Regency), Tanah Lot (Tabanan Regency) and Seririt (Buleleng Regency) (BPBD, 2017a). The early warning system is connected to the data centre of the Geophysics, Climatology, and Meteorology Agency of Bali to allow monitoring of tsunami risk, provide a basis for public announcements on the level of risk; and coordinate with the BPBD to decide any emergency actions. The devices are regularly maintained by the BPBD with a monthly test occurring on the 26th day of each month to ensure all sirens work well and are able to submit data on any tsunami potency to the Geophysics, Climatology, and Meteorology Agency.

The BPBD has also developed an information dissemination system that uses a number of media, including VHF Radio, which is connected to all villages in Bali, social media, email, television, relevant websites, text messaging and a call centre. This system also aids in educating and increasing awareness of local people on the risks of earthquake and tsunami. The mitigation program does this by promoting understanding on earthquake and tsunami disasters and their dangers; and also widening knowledge on what to do and where to escape to when such natural disasters happen.

To support mitigation planning, Bali's BPBD has built a temporary evacuation centre on the isle of Serangan, a small village located in southern Denpasar (Plate 4.2). This three-storey building accommodates 2000 people in the matter of emergency. This centre is completely furnished with and equipped with kitchen equipment to meet the immediate needs of disaster refugees. The development of this evacuation centre was initiated by research conducted by the agency which concluded that evacuation in this small island required vertical evacuation as it is impossible to escape from tsunami disaster horizontally within 15 minutes due to its location by the ocean. Other flat areas in the southern coastal regions such as Kuta, Sanur and Nusa Dua use horizontal evacuations which allow people on the seashore to escape. As the evacuation centre numbers are still very limited, it will eventually be extended to other villages along the southern coast of Bali that are assumed to be vulnerable to tsunami.

The BPBD has also established a disaster prepared village (*desa siaga bencana*) program which aims to develop resilient disaster villages in which local people have been trained with respect to earthquake and tsunami disaster risks in conjunction with officials prepared by the agency. This program has been applied in villages located in the southern coastal regions of Badung regency such as Tanjung Benoa, Nusa Dua, Jimbaran, Kuta and Legian villages; and also in Denpasar city, including

Mitigating Earthquake and Tsunami Risks in Coastal Tourism Sites in Bali 75

Plate 4.2 Temporary Tsunami Evacuation Centre in Serangan Village
Source: Author.

Sanur and Serangan villages, where most of hotel resorts and tourist attractions are located. In addition to training programs for local people, the Regional BPBD has also developed 'train the trainer programs' in which officials required to attend short courses on disaster mitigation conducted in Bali, the central government of Jakarta, and even overseas in order to increase their disaster management competences. In 2017 for instance, two officials of the agency were invited to attend a two-week short course on disaster risk mitigation offered by the National Critical Care and Trauma Response Centre (NCCTRC) of the Northern Territory Department of Health in Australia. Three months after the course, a team from the Australian department came to Bali and trained all officials working for the Regional Disaster Mitigation of Bali in a Major Incident Medical Management and Support Course. The training included an in-class course to increase knowledge of disaster and risk mitigation and also an out of class simulation to allow staff to practice the knowledge gained during the course and acting out their disaster management roles (BPBD, 2017b; Bali Post, 2017). This government to government program is not only aimed at increasing the competence of the BPBD officials on disaster management, but also strengthening bilateral cooperation between Indonesia and Australia.

Immediate and Post-Disaster Mitigation Programs

Emergency response management deals with impacts of the immediate disaster and includes searching, rescuing and evacuating victims; fulfilling the basic needs of the victims; organizing and managing displaced persons; and recovering the affected infrastructure. The earthquake and tsunami disaster mitigation programs organized by the BPBD since 2012 have not been tested in a real case. However, the planned programs have been frequently applied to other natural disasters occurring on the island including floods, landslides and also the 2017 volcanic eruptions which continue to the present day.

In term of the post-disaster management plan, there are two major programs included in the long-term mitigation and recovery program including building reconstruction and the rehabilitation of people affected by disasters. The reconstruction of public service facilities and housing involves the Bali Department of General Works and regional government. Rehabilitation programs dealing with stress and trauma are undertaken in cooperation with the Health Department of Bali which coordinates the regional public hospitals and other medical centres located in each district. Funding for these programs comes from central, provincial and regional governments; voluntary donations; and international aid. Recovery programs are focused on local people and public services, as well as tourism amenities and disaster affected tourists. As Huang and Min (2002) argue, tourism recovery is essential and aimed at attracting the tourists back to visit the destination and increasing the number of tourist visits.

In 2014 the Bali BPBD also formulated a strategic plan which serves as the principal guidance for mitigating the impacts of earthquake and tsunami disasters in Bali. This plan is aimed at ensuring integration among the planning, budgeting, application and supervisory functions; prescribing the vision, mission, program and activities; executing the annual plan for a five-year term; and stipulating policy and programs for in accordance with the principal duties and function of the agency.

Stakeholder involvement in disaster mitigation

Mitigating the risks of natural disasters like earthquake and tsunami requires the integrated involvement of stakeholders. The government of Bali through the BPBD cooperates with numerous stakeholders including all government bodies (regency, sub-district and village), local people, hoteliers and other tourism related industries to work together to reach the maximum preparedness for dealing with disasters. The level of preparedness of particular tourism stakeholders is different from each other and can require education to help develop competence for disaster mitigation (Muttarak & Pothisiri, 2013).

Accordingly, the agency has developed a number of programs to encourage the participation of stakeholders to minimize the risks of earthquake and tsunami (BPBD, 2017a). These cover familiarization with the dangers to local people residing in the southern coastal region where tsunami potentially occurs, e.g. in Tanjung Benoa, Nusa Dua, Kuta, Serangan and Sanur villages. Local communities are invited to *Bale Banjar* (a sub-village meeting pavilion) and receive material on earthquake and tsunami and their dangers by the assigned government officials, including information on the detailed maps and evacuation zones available for their villages. They are also obligated to be involved in a simulation to provide a brief overview as to how disaster can be mitigated. These programs are conducted regularly each year moving from one village to another to ensure all villages along the coastal zones are well trained.

The involvement of local people on tsunami disaster mitigation simulation is significant. It not only encourages them to engage with training, but also includes 'their culture' especially relating to traditional technology such as the *Kul Kul,* a wooden bell primarily used for traditional and religious-related activities. This wooden bell is found in all traditional village meeting pavilions and Hindu temples in Bali. Balinese people mainly use Kul Kul to invite the members of a village to gather in the Bale Banjar to attend a meeting or to be involved in *ngayah* – a voluntarily devotional service conducted in the temples. In some cases, Kul Kul have been used to alert Balinese people in the case of emergency. During the Mount Agung Eruption for instance, these bells were used during the evacuation process of people living within one to twelve kilometres from the mountain. The author's family and relatives who live nine kilometres from Mount Agung never took notice of the village officials that approached them or the miscellaneous messages circulated by their family and friends to vacate their houses before the alert status in November 2017 as they considered that it was not about to erupt based on their experiences of the 1963 eruption. However, when the Kul Kul was hit and sounding in all meeting pavilions; they left their houses voluntarily and very promptly gathered in Bale Banjar which served as assemble points for evacuation (author's field note 22 November 2017). This suggests that Kul Kul have such a significant power in traditional Balinese communities that it is used both for traditional and Hindu related-rite activities as well as for emergency alerts. For these reasons, the use of the Kul Kul has been codified and agreed with Balinese communities to be used to provide a tsunami alert to support the modern early warning system devices installed in the coastal zones. This also assists Balinese people in being able to sustain their culture even if the island has become 'globalized' due to the rapid growth of domestic and international tourist visits (Subadra, 2015).

Familiarization and simulation programs have also been introduced to employees of hotels and restaurants sited within the tsunami zones. Such information cannot only inform them how to save their own lives but also the best procedures to evacuate guests. Hotels that had undertaken such programs can be certified by the BPBD in terms of earthquake and tsunami disaster preparedness. These simulation and certification programs are organized in cooperation between the Bali BPBD, the Bali hotel association and the association of Indonesian hotels and restaurants.

Such projects correspond to the United Nations (UN) (2007) 'Bali Declaration on Sustainable Tourism Development' whereby hotels and other tourism-related businesses are encouraged to become competent on disaster crisis and risk management. Safety and risk management programs have also been included in certification programs developed as part of national policy under the *Tourism Act of Republic of Indonesia Number 10 of 2009* and also regional regulations in Bali which obligate hotels to accomplish compulsory certification as part of their business permit requirements. As of early 2018, hotels with 43 stars have been assessed and certified to be well-prepared-disaster hotels. The online publication of this certified assessment on the hotel's website helps consumers find safe and secure accommodation for their holidays in Bali and also serves as an 'added value' in supporting the marketing of hotel rooms.

The Bali BPBD has also facilitated cooperation between traditional villages and hotel management to use the nearest multi-storey hotels as assembly points and evacuation centres for local people when there is a tsunami. This applies the same notion of vertical evacuation, as on Serangan island, since the area of Tanjung Benoa has a flat foreland where hundreds of hotels and other tourism facilities are situated.

Currently, there are at least three 5-star hotels officially designated by the government as assembly points in case of an earthquake or tsunami emergency. The people living near to the hotels have been trained by the assigned officials to recognize the nearest places to assemble when a tsunami occurs. The involvement of the hotel sector in disaster management has a great meaning in Bali in terms of the prevention and mitigation of natural disasters.

Conclusion

The popularity of Bali as a tourism destination means that it attracts millions of international tourists each year. and serves as the economic engine of the island. The southern coastal regions of Sanur, Kuta and Nusa Dua are currently the main tourist hubs where most hotels and resorts have been established. However, these locations, along with other parts of the island, are at great risk from earthquakes and tsunami. In other words, Bali is unavoidably susceptible to the threats of those disasters which can damage local people's and tourism properties and kill local residents and tourists in the coastal region. The only efforts which can be prepared to prevent and mitigate such disasters are increasing the existing disaster management capacities including improvements in preparedness and levels of knowledge of locals and tourism employees; improving tourist awareness of appropriate actions when a disaster occurs; upgrading the competencies of the government officials who deals with disaster mitigations through training, workshop and seminars; improving coordination within and between government departments and tourism stakeholders; and finally, increasing the budgets for disaster mitigation at regional, provincial and central government levels.

References

Badan Meteorologi Klimatologi dan Geofisika. (2017) *Gempabumi Terkini* [online]. See http://www.bmkg.go.id/gempabumi/gempabumi-terkini.bmkg.

Badan Penanggulangan Bencana Daerah (BPBD) (2017a) Renstra Badan Penanggulangan Bencana Daerah Provinsi Bali Tahun 2014-2018 [Bali Provincial Regional Disaster Management Strategic Plan Years 2014-2018] [online]. See http://bpbd.baliprov.go.id/?page=Renstra-Badan-Penanggulangan-Bencana-Daerah-Provinsi-Bali-Tahun-2014-2018&language=id&domain=

Badan Penanggulangan Bencana Daerah (BPBD) (2017b) *Pelatihan 'Major Incident Medical Management and Support Course - Advanced (MIMMS)'* [online]. See http://balisafety.baliprov.go.id/berita/pelatihan-major-incident-medical-management-and-support-course---advanced-mimms.html.

Badan Pusat Statistik Provinsi Bali (Statistics of Bali Province) (2019a) Jumlah Wisatawan Asing ke Indonesia dan Bali, 1969–2018 (Number of Foreign Visitor to Indonesia and Bali, 1969–2018). Denpasar: Author.

Badan Pusat Statistik Provinsi Bali (Statistics of Bali Province) (2019b) Jumlah Kamar pada Hotel Bintang, Number of Rooms Available of Classified Hotels 2000–2018. Denpasar: Author.

Badan Pusat Statistik Provinsi Bali (Statistics of Bali Province) (2019c) Jumlah Kamar pada Hotel Non Bintang dan Akomodasi Lainnya, Number of Rooms Available of Nonclassified Hotels and Other Accommodations 2000-2018. Denpasar: Author.

Bali Post. (2017) *NCCTRC Australia Latih Tim Penanggulangan Bencana Bali* [online]. See http://www.balipost.com/news/2017/05/31/10171/NCCTRC-Australia-Latih-Tim-Penanggulangan.html.

Garg, A. (2015) Travel risks vs tourist decision making: A tourist perspective. *International Journal of Hospitality & Tourism Systems* 8 (1), 1–9.

Glaesser, D. (2003) *Crisis Management in the Tourism Industry*. Oxford: Butterworth-Heinemann.

Hodgkin, D. (2016) *Emergency Response Preparedness in Indonesia: A Consultation Report prepared exclusively for the Indonesia Humanitarian Country Team*. Humanitarian Benchmark Consulting [online]. See https://www.who.int/docs/default-source/searo/indonesia/non-who-publications/2016-emergency-response-preparedness-report-in-indonesia-eng.pdf?sfvrsn=1905f2b4_2.

Huang, J.H. and Min, C.H.J. (2002) Earthquake devastation and recovery in tourism: The Taiwan case. *Tourism Management* 23 (2), 145–154.

Jensen, S. and Svendsen, G.T. (2016) Social trust, safety and the choice of tourist destination. *Business and Management Horizons* 4 (1), 1–9.

Kelman, I., Spence, R. and Palmer, J. (2008) Tourists and disasters: Lessons from the 26 December 2004 tsunamis. *Journal of Coastal Conservation* 12 (3), 105–113.

Khomarudin, M.R., Strunz, G., Ludwig, R., Zoßeder, K., Post, J., Kongko, W. and Pranowo, W.S. (2010) Hazard analysis and estimation of people exposure as contribution to tsunami risk assessment in the west coast of Sumatra, the south coast of Java and Bali. *Zeitschrift für Geomorphologie, Supplementary Issues* 54 (3), 337–356.

McCool, B.N. (2012) The need to be prepared: Disaster management in the hospitality industry. *Journal of Business & Hotel Management* 1 (1), 1–5.

Moore, J. (1970) *Bali, Island of a Thousand Temples*. Singapore: Asia Pacific Press.

Muttarak, R. and Pothisiri, W. (2013) The role of education on disaster preparedness: Case study of 2012 Indian Ocean earthquakes on Thailand's Andaman Coast. *Ecology and Society* 18 (4), 1–16.

Nguyen, N., Griffin, J., Cipta, A. and Cummins, P. (2015) *Indonesia's Historical Earthquakes: Modelled Examples for Improving The National Hazard Map*. Canberra: Geoscience Australia.

Nguyen, D., Imamura, F. and Iuchi K (2016) Disaster management in coastal tourism destinations: The case for transactive planning and social learning. *International Review for Spatial Planning and Sustainable Development* 4 (2), 3–17.

Picard, M. (1996) *Bali: Cultural Tourism and Touristic Culture*. Singapore: Archipelago Press.

Powell, H. (1982) *The Last Paradise*. Kuala Lumpur: Oxford University Press.

Strunz, G., Post, J., Zosseder, K., Wegscheider, S., Mück, M., Riedlinger, T., Mehl, H., Dech, S., Birkmann, J., Gebert, N. and Harjono, H. (2011) Tsunami risk assessment in Indonesia. *Natural Hazards and Earth System Sciences* 11 (1), 67–82.

Subadra, I.N. (2015) Preserving the Sanctity of Temple Sites in Bali: Challenges from Tourism. Unpublished PhD. University of Lincoln.

Suppasri, A., Futami, T., Tabuchi, S. and Imamura, F. (2012) Mapping of historical tsunamis in the Indian and Southwest Pacific Oceans. *International Journal of Disaster Risk Reduction* 1, 62–71.

Syahbana, D.K., Kasbani, K., Suantika, G., Prambada, O., Andreas, A.S., Saing, U.B., Kunrat, S.L., Andreastuti, S., Martanto, M., Kriswati, E. and Suparman, Y. (2019) The 2017–19 activity at Mount Agung in Bali (Indonesia): Intense unrest, monitoring, crisis response, evacuation, and eruption. *Scientific Reports* 9 (1), 1–17.

United Nations (UN) (2007) Bali declaration on sustainable tourism development: Plan of action for sustainable tourism development in Asia And the Pacific, Phase II (2006–2012) and regional action programme for sustainable tourism development. *ESCAP Tourism Review* 26. New York: UN.

Utomo, Y.W. (2011) Sejarah Gempa dan Tsunami di Bali. *Kompas*, 13 October [online]. See at http://sains.kompas.com/read/2011/10/13/21102227/Sejarah.Gempa.dan.Tsunami.di.Bali.

Withnall, A. (2016) The Best Islands in the world in 2016. *Independent*, 24 April [online]. See http://www.independent.co.uk/travel/news-and-advice/tripadvisor-best-islands-in-the-world-2016-a6998546.html#gallery.

5 It is Not Just About a Convention Centre: Expectations and Disillusions from Tourism-Relevant Stakeholders in Post-Earthquake Christchurch

Alberto Amore

Introduction

In 2010 and 2011, the city of Christchurch, New Zealand was severely hit by a series of earthquakes that damaged most of the city centre. The local tourism and hospitality industry were among the sectors that were most affected, with the demolition or closure of key attractions and amenities in the city. Overall, the earthquakes had a negative impact on the appeal of Christchurch and the wider Canterbury region as a place to visit between February 2011 and late 2014 (Hall *et al.*, 2016).

Following the February 2011 earthquake, tourism-relevant stakeholders brought the issue of tourist spaces recovery into discussion with the leading recovery authorities. During the drafting and launch of the city centre rebuild plan, a narrow group of representatives from tourism and hospitality urged local and national authorities to quickly re-establish core tourism services in the city centre (Christchurch & Canterbury Tourism (C&CT) & Tourism Industry Association of New Zealand (TIANZ), 2011; Christchurch City Council (CCC), 2011a, 2011b). However, their insistances steadily fell short following the establishment of the Christchurch Central Development Unit (CCDU) and the release of the central city recovery plan in July 2012 (CCDU, 2012a). As the rebuild went on, the shaping of the city and of its major amenities suggest that the approach to recovery with regards to tourism and hospitality was piecemeal and characterized by a series of roadblocks, metagovernance failures and clientelisms. The slow recovery of Christchurch had a negative impact in terms of tourist arrivals and nights, with the city performing poorly compared to the rest of New Zealand (Canterbury Development Corporation (CDC), 2014). Data from 2013 and 2014, in particular, suggest that the tourism economy of Christchurch performed below the worst-case scenario reported in the *Greater Christchurch Visitor Recovery Plan* (C&CT, 2012a).

The following chapter presents findings from extensive research undertaken in Christchurch. The chapter illustrates the crisis and stagnation of the visitor economy in Christchurch in the period between 2010 and 2016. Findings from documents and interviews with relevant stakeholders are used to analyse episodes of governance focusing on urban redevelopment projects relevant to the local tourism and hospitality sectors. The chapter deploys a longitudinal, socio-interpretive approach to illustrate expectations and disillusions from tourism-relevant stakeholders in post-earthquake Christchurch.

Literature review

Research focusing on representations and views of tourism stakeholders gained momentum towards the end of the 1990s (Jamal & Getz, 1999; Yuksel *et al.*, 1999). The focus on stakeholders, sheds light on the 'plurality of organizational interest groups and the political nature of organizational goal setting and policy implementation' (Treuren & Lane, 2003: 4). Nonetheless, the shift from macro-level structures to micro-level situated social practices only occurred in the mid-2000s, with findings from policymakers (e.g. Stevenson *et al.*, 2008), local representatives of the tourism industry (e.g. Bramwell & Meyer, 2007), event organizations (e.g. Dredge & Whitford, 2011) and the local community (e.g. Truong *et al.*, 2014).

According to McKercher (1999: 425), the occurrence of crises and disasters in tourism destinations is a marked example of 'the chaotic nature of tourism systems'. The exposure of destinations to disturbances can vary depending on their intensity (Becken, 2013; Biggs *et al.*, 2012; Calgaro *et al.*, 2013; Lew, 2014). Undoubtedly, 'people perceive and manage slow changes in the environment, culture and society in a different manner than they do under sudden major shocks to these systems' (Lew, 2014: 17). Moreover, there are different wants and priorities when coping with crises and disasters between the tourism industry stakeholders and the wider community (Evans & Elphick, 2005; Lew, 2014; Ritchie *et al.*, 2014). Finally, priorities and needs of tourism stakeholders at large change throughout the different phases of crisis and disaster (Faulkner & Vikulov, 2001). For instance, the priorities for hotel owners forced to closure in the aftermath of a disaster differ substantially from those common to the emergency phase (Gurtner, 2007).

'The study of post-disaster recovery is in its infancy and there is as yet no body of theory to guide researchers' (Olshansky *et al.*, 2012: 173). This is particularly true in the study of post-disaster and recovery governance of tourist areas. For the purposes of this chapter, the focus is on recovery phase, which begins when the immediate emergency ends and can take years to resolve. It is in this key phase that, under extraordinary time constraints, recovery is framed by the discourses of different interests over the best solutions for redevelopment (Amore & Hall 2016a; Gotham & Greenberg, 2014).

The current literature acknowledges the phase of recovery as consisting of reconstruction and reassessment of tourism destinations, but eventually overemphasizes on marketing and re-branding strategies (Carlsen & Liburd, 2008; Orchiston & Higham, 2016). Nonetheless, the recent call for a social sciences insight looking at the relationships between crises and socioeconomic institutions (Hall, 2013), suggests that current research in tourism, crises and disasters is moving towards a social theory approach advocated in tourism planning and governance (Bramwell, 2011).

Tourism-relevant stakeholders and post-disaster metagovernance

To date, only a handful of works address post-disaster metagovernance of destinations from the perspective of tourism stakeholders. Scott *et al.* (2008: 3) acknowledge that the 'effects of a crisis may be transferred across system boundaries by organizational relationships' and thus pave the way to new, more efficient forms of destination networking . Evidence from post-earthquake Christchurch, shows that 'other broader industry networks have developed as a consequence of the earthquakes' (Orchiston & Higham, 2016: 81). These works, however, overlook how metagovernance structures change following a triggering event (Jessop, 2011; Johnson & Mamula-Seadon, 2014).

Post-disaster recovery is seen as a window of opportunity to create incentives for the attraction of international capital for the rebuild and rebrand destinations to international tourism. Evidence from Sri Lanka, Thailand, Honduras and the Unites States suggest that government stakeholders look at international tourism as an opportunity to generate foreign revenue and boost tourism (Gotham & Greenberg, 2014; Gunewardena, 2008; Stonich, 2008). In particular, the re-branding of New York and New Orleans following 9/11 and Hurricane Katrina are often mentioned by government stakeholders 'to teach best practices for market-oriented growth, while elevating the status of cities for the purpose of interurban competition' (Gotham & Greenberg, 2014: 221).

Government and key tourism industry stakeholder legitimize the rhetoric from international best-case studies in post-disaster tourism recovery by side-lining the instances of other stakeholders. Findings from Cyprus following the 2012-2013 sovereign debt crisis suggest that the hospitality sector determined what were the 'legitimate values in tourism development' (Farmaki *et al.*, 2015: 183). Similarly, Thai government authorities overrode local practices for tourism development in post-tsunami Thailand, with a top-down recovery strategy based on international best-case studies that overlooked the potentialities of the local community in the redevelopment process (Larsen *et al.*, 2011).

Particularly for urban tourism contexts, the simultaneous co-presence of different post-disaster needs (e.g. business recovery, infrastructure redevelopment) further complicates the assessment of the context and of the different interests and values involved. In post-disaster contexts the 'political conflicts over redevelopment are deeply spatialized as people negotiate and renegotiate meanings of and control over urban space' (Gotham & Greenberg, 2014: 95). On the one hand, there is a greater citizenry engagement and proactive input on urban redevelopment. On the other hand, we have growth coalitions comprising big private interests and a quest for large-scale redevelopments (Olshansky *et al.*, 2012).

Evidence from New York and New Orleans (United States) and Arugam Bay (Sri Lanka) suggests that lobbying groups divert resources and decisions to matters dear to their personal gain. Disasters, in turn, serve as pretext to clear areas and support public/private redevelopment partnerships that please the interests of corporations while further increasing the vulnerability of the people affected by disasters (Gotham & Greenberg, 2014; Johnson & Olshansky, 2010). Disasters become fertile ground for radical institutional arrangements among a narrow group of stakeholders that determine 'the level of government that controls the flow of money and how it is acquired, allocated, disbursed, and audited' (Johnson & Mamula-Seadon, 2014: 596).

This chapter looks at wants, needs, beliefs and interests of tourism-relevant stakeholders on the redevelopment of Christchurch as urban destinations in the aftermath of the 2010 and 2011 Canterbury earthquakes. The chapter embraces the notion of metagovernance in tourism (Amore & Hall, 2016b) in acknowledgement of the policy failures and re-organization of tourism destinations in post-disaster contexts. The chapter applies insights from spatial planning theory and deploys a narrative approach for the analysis and explanation of governance episodes already in use in tourism policy and planning (Hall, 2008).

Methodology

The findings reported in this chapter are part of an extensive research undertaken between 2013 and 2015. Reports and strategy documents were collected from the Canterbury Earthquake

Recovery Authority (CERA), the Christchurch City Council (CCC), the Department of the Prime Minister and Cabinet (DPMC), the New Zealand Parliament Library, Christchurch & Canterbury Tourism (C&CT) and the Tourism Industry Association of New Zealand (TIANZ). Data for the analysis was collected from key tourism stakeholders and government representatives leading the recovery of Christchurch between 2010 and 2015. A two-round series of semi-structured interviews with 18 tourism-relevant participants was undertaken between May and November 2015. The researcher used both direct recruiting and snowballing in the selection of participants from different spheres of the government, economy and civic society involved at some stage with the recovery. For confidentiality purposes, this chapter does not report the organizational affiliations of participants and uses fictional names.

Documents and interview extracts were coded with the support of NVivo and grouped under pre-established narratives. The analysis sought to assess the awareness among tourism-relevant stakeholders in relation to the recovery of Christchurch and its core tourist areas. The focus on awareness among tourism-relevant stakeholders underpins previous research in central city regeneration governance (Healey et al., 2003). For each of the main episodes of metagovernance, the analysis looked at knowledge, mobilization and relational resources (Healey et al., 2003) from the perspective of urban tourism redevelopment.

Context: Christchurch, New Zealand

The city of Christchurch is the major urban area of the South Island of New Zealand and is located in a region prone to medium-to-high intensity earthquakes, liquefaction and lateral spreading (Cubrinovski & McCahon, 2012; Orchiston, 2012). Most of the damage from the earthquakes in 2010 and 2011 occurred in the Christchurch central city area, a relatively small yet important economic cluster of commercial, financial and tourism-related activities, as well as in suburbs along reclaimed swampland and the coast (Hall et al., 2016). Estimates on the total cost of damage caused by the earthquakes varied between NZD$ 30 billion and NZD$ 40 billion (New Zealand Treasury (NZT), 2013), while the rebuild costs are estimated as high as NZD$ 40 billion (Reserve Bank of New Zealand (RBNZ), 2016). As at the time of this book being published, some 10 years after the first earthquake, the rebuild is still occurring and final costs are yet to be determined.

Before the earthquakes, the city of Christchurch was home to around 16% of the total tourism activity of New Zealand and 10.2% of the total tourism spending (New Zealand Ministry of Business, Innovation and Employment (NZMBIE), 2010; Orchiston et al., 2012). The repeated aftershocks, adverse media coverage, the relevant infrastructural damage in many facilities, the slow pace of recovery and the issues with settlement claims with insurance made the recovery process very painful to the tourism-relevant stakeholders (Orchiston et al., 2012). Some six years after the major earthquakes, the climate of uncertainty among tourism-relevant stakeholders still remained high.

The redevelopment of Christchurch was far from straightforward. Leading recovery authorities sought to rebuild the city based on international best practice urban design (DPMC, 2012). However, national and local authorities were at odds with respect to the rebuild strategy of the city. The National Government and the officials of CERA and the CCDU identified anchor projects as the preferred urban redevelopment strategy (CCDU, 2012b), while the CCC envisioned the creation of a distinctive central city consisting of low-rise urban precincts acting as clusters for organic growth (CCC, 2011a). The mandate to CERA and the CCDU to lead the rebuild of Christchurch and its central city was the result of a series of Cabinet decisions, as foreseen in the dedicated earthquake recovery legislation.

The recovery strategies and the redevelopment projects carried throughout the lifespan of CERA (2011–2016) exemplified the disaster capitalism paradigm seen in other cities affected by natural hazards (Gotham & Greenberg, 2014). In particular, the national recovery authorities saw the recovery 'as a once in a lifetime opportunity to radically change the highly parcelled ownership of land in the CBD and sell allotments to attract major international developers' (Amore & Hall, 2016a: 190). This, in turn, raises questions as to whose interests the recovery authorities looked after in the drafting of the redevelopment strategy of Christchurch.

The implementation of the recovery strategy and the delivery of the anchor projects fell relatively short. The delays with the site projects and the allocation of resources for the rebuild cooled the interest of international investors, while the local developers encountered enormous difficulty in complying with the financial conditions to meet the rebuilding requirements (McCrone, 2014). The partnership between the Carter Group and Plenary Group for the Convention Centre Precinct announced in August 2014 never came to fruition (McDonald, 2016). The delivery of the new Stadium was postponed to 2022 due to issues with the insurance settlement and the lack of a definitive business case (Amore & Hall, 2017). Finally, the uncertainties over the future of the Christchurch Cathedral and the Square were only dissipated at the end of 2017, with the announcement of a NZD$ 10 million grant towards the reinstatement of the building (CCC, 2017).

Findings

Values and beliefs among stakeholders

Tourism-relevant stakeholders had diverging views with regards to the recovery of Christchurch through tourism. The instances of hospitality and tourism businesses were at odds with those drafting the economic recovery strategy for Christchurch. The latter advocated for a shift from tourism to the visitor economy at large (CDC, 2014), but leading tourism stakeholders were more focused on hotel development (*Blair* and *Ted*). This, in turn, contributed to a sectorial appraisal of tourism within the wider recovery framework. As one participant stressed

> I think it's people's attitudes towards tourism and hospitality. They think: 'well, there are more important things' or whatever [...] But tourism is an important part of our economy! [*Jeb*]

Undoubtedly, there were differences in views with respect to tourism and its role in the recovery of Christchurch among relevant stakeholders. For instance, *Frank* saw 'opportunities for developers related to tourism, retail and shops' coming from projects outlined in the Christchurch recovery plans. However, *Gabriel* argued that the impact of the recovery projects for leisure and tourism would have been 'pretty minimal'. A third participant highlighted that there was 'no discussion about tourism, the value of tourism in the city and the indirect effects' (*Edward*) in the recovery strategy and plans for Christchurch and its city centre.

Tourist operators in Christchurch and New Zealand expressed concern about the loss of facilities and amenities in the city and the impact on the destination appeal among international tourists (Wallace & Simmons, 2012). One of the participants advocated for the creation of 'corridors of neutrality' for tourists and tourism amenities within the city (*Joseph*). Tourism-relevant stakeholders stressed the importance of retaining key attractions in the city (*Elton, Frank, Kevin* and *Joseph*). However, there were diverging opinions and interests between the leading recovery authorities and the tourism-related stakeholders when it came to the implementation of the recovery plan. The retention of relevant heritage attractions in the city was not part of the recovery agenda. As one participant stressed with respect to the Christchurch Cathedral:

> [...] that should have been a focus. Because that's a place that people – I think – think a lot about. And there have been lots of plans for what to do with the square. But with four years on and more nothing has happened. It is still like it was, virtually, the day after the earthquake. [*Elton*]

Tourism-related stakeholders stressed the importance of replacing demolished amenities to increase the appeal of the city following the earthquakes. The Minister for the Canterbury Earthquake Recovery, CERA and the CCDU were adamant in putting forward their anchor project-driven agenda (CCDU, 2012b). This was the case for the Convention Centre Precinct, with government and private developers reiterating the benefits for large-scale venues for the lucrative business tourism market (Convention & Incentive New Zealand (CINZ), 2014). The C&CT CEO, Tim Hunter, publicly highlighted the pros of the new conference facility by stressing the benefits for the tourism economy at large (Hunter, 2015). Participants reiterated the positions of the recovery authorities on the need for a good-sized, international conference venue in the heart of the city and the ripple effect it generated in terms of hotel development (*Elton, Sam* and *Thomas*). Similarly, the *Greater Christchurch Visitor Recovery Plan* considered the Convention Centre Precinct as 'essential to re-instate high value business tourism in a city that has previously excelled in the conference and convention sector' (C&CT, 2012a: 7). One participant stressed the centrality of the Convention Centre Precinct in the tourist economy of the city when talking about the nearby performing arts district:

> We liked the area was going to be close to the Convention Centre. So, people from the convention would logically have all that stuff going on in that part of town. [*Gabriel*]

The loss of sporting and other events were often cited as detrimental for the recovery of Christchurch as a tourist destination. Tourism-relevant stakeholders acknowledged how events could 'create an enormous amount of revenue for the city in terms of accommodation, hotels, hospitality, food and drink and entertainment' (*Frank*). One participant provided evidence from the Foo Fighters concert held in Christchurch in February 2015, which generated a direct economic turnaround of NZD$ 4.75 million and argued that:

> Christchurch has missed out on a dozen of these events over the last year, year and a half, maybe two years. And that's a pretty substantial amount of benefit the city has lost out on. [*Richard*]

Tourism-relevant stakeholders stressed the need for Christchurch to invest on brand new event facilities. Three participants argued that building a new stadium was extremely important and that the option for a covered stadium would benefit the hosting of big concerts and shows (*Sam, Thomas* and *Richard*). One of them described the project for the new 35,000-seat covered stadium in the Christchurch city centre as 'a catalyst project that massively enhanced the hospitality and tourism industry' (*Thomas*). The Canterbury Rugby Union, and the key recovery authorities claimed that a city like Christchurch needed a large-scale stadium for hallmark sports events like the All Blacks test matches and the Lions Tour (*Colin, Richard* and *Sam*).

Not all tourism-relevant stakeholders agreed on the type, size and priority of projects to boost tourism in the city. Particularly with the Convention Centre Precinct, stakeholders raised concerns regarding the appeal Christchurch has as a conference destination following the earthquakes (*Elton*) and the decrease of interest among hospitality investors due to the delays in the project (*Sam*). With regards to the stadium, participants stressed that the temporary stadium in Addington was enough to meet the demand for sports events in Christchurch (*Adam* and *Elton*) and that the proposed 35,000-seat option was uneconomic (*Adam, Gabriel, Roy*). Unsurprisingly, relevant stakeholders like the CCC and the C&CT considered the building of the new stadium as a low-priority and rather advocated for the building of other sports attractions to rebrand Christchurch as a sports tourism destination (C&CT, 2012a; CCC & C&CT, 2013).

The emphasis on anchor projects among key recovery authorities and leading tourism business stakeholders was fiercely criticized by other tourism-relevant stakeholders:

> If you talk to people in [the] tourism industry and business people they'll say Convention Centre. [...] If you talk to developers around here, they go: 'Convention Centre'. If you ask what you need first, they say: 'Convention Centre'. [*Blair*]

> Overseas there are hundreds of Convention Centres. Hotel developers won't start building until the Convention Centre starts because you have to market your conventions. [...] Hotel developers are just on hold until this happens. [*Frank*]

At the time of fieldwork (late 2015), there was substantial concern over the stage of recovery among tourism-relevant stakeholders (*Joseph*). Prior to the earthquakes, the city role as a gateway to the South Island meant that the length of stay among international tourists was comparatively lower than in other major cities in Australia and New Zealand (*Edward* and *Elton*). Participants argued that recovery authorities missed a once in a lifetime opportunity to reinvent Christchurch as a tourist city through iconic attractions and facilities (*Edward*). As participants emphatically stated:

> What this Government fails to understand – or to consider – is that the built environment is of interest to people. You don't go on a plane and go on a holiday because you want to sleep in a room. You want to go [to] a place because it has attractions. [*Edward*]

One of the things that one organization said to us was: 'Well, where is your Sydney Harbour Bridge? Where is your Sydney Opera House?' I think that's what Christchurch currently lacks. That is, facilities that are going to draw people here and keep them here. We lost an awful lot of visitors [and] we desperately need facilities that are going to keep people here. [But] I don't think we really addressed that yet. [*Elton*]

Degree of collaboration among stakeholders

Dedicated tourism stakeholders and organizations played only a marginal role in the process of planning decision making that eventually culminated with the Christchurch recovery strategy and the central city rebuild plan (*Gabriel*). With regards to the CCC's central city plan, Tim Hunter and representatives from C&CT and TIANZ only provided written feedback following the release of the first draft in August 2011 (C&CT & TIANZ, 2011). Following the establishment of the CCDU in April 2012, managers and consultants of the C&CT began working with CERA on the *Tourism Recovery Plan* (*Elton* and *Jeb*). However, CERA never put in place the plan (*Jeb*). Rather, recovery authorities shifted the focus on the Convention Centre Precinct and the need to expand the hotel room capacity of Christchurch (*Elton* and *Sam*).

According to participants, people working for CERA and CCDU and the hierarchical structure around the Minister for the Canterbury Earthquake Recovery were adverse factors for the recovery of tourism in Christchurch (*Elton, Jeb* and *Ted*). Two participants explicitly referred to the Minister for the Canterbury Earthquake Recovery, Hon. Gerry Brownlee, as a 'major impediment' (*Jeb*) and a 'bully' reluctant to work with the community (*Elton*). Monthly meetings with CERA and CCDU officials between 2011 and 2012, 'led to very little outcomes for the tourism sector' (*Elton*). Since then, the recovery authorities progressively side-lined key tourism stakeholders and pursued its anchor project-driven agenda (*Elton*). This raised eyebrows as to whether the metagovernance of post-earthquake Christchurch was harmful for tourism:

With the benefit of hindsight, I do wonder whether the Government made the right decision. I suppose they had to, from a financial point of view, but I wonder whether they should have looked at empowering existing organizations rather than establishing another one. [*Jeb*]

The role of tourism-relevant stakeholders and the importance given to organic tourism development was very limited. Projects submitted in 2011 such as the tourism precinct and the River of Arts (Arts Voice, 2011) were not considered in the central city plan, while dedicated tourism authorities such as the C&CT were only involved with the promotion and re-branding of Christchurch (C&CT, 2012b, 2013).

Similarly, private companies running key tourist services like punting on the Avon River, the Christchurch Tramway and the Christchurch Gondola were relegated far back in the decision-making process for the recovery of the city as a tourist destination. More importantly, the CERA and the CCDU stripped the local Canterbury Development Corporation (CDC) of the mandate to develop the economic strategy and attract tourism private investors to Christchurch (CCDU, 2013). These episodes, in turn, generated 'a sense of frustration' among Christchurch's tourism-relevant stakeholders (*Ted*).

Two emblematic episodes best illustrate the engagement of recovery authorities with tourism-dedicated stakeholders. On the one hand, CERA's Tourism Manager focused exclusively on monitoring tourist arrivals and tourist nights in the city (*Elton* and *Jeb*). Most importantly, he had no advisory role on the drafting of the central city recovery plan (*Joseph* and *Jeb*). On the other hand, the engagement of CERA over the Eden Project was far from being collaborative. As one of the participants acknowledged, CERA staff turned the entire process 'extremely frustrating' (*Elton*) and it took nearly three years to come with a preliminary proposal for the project (C&CT, 2015).

Despite the frictions with CERA and the CCDU, tourism-relevant stakeholders were able to embark on a range of marketing initiatives to re-brand Christchurch. The Ministry of Business, Innovation and Employment funded a joint marketing campaign in partnership with C&CT, the CCC and the Christchurch International Airport (NZMBIE, 2015) as well as the *Tourism Recovery Plan* submitted to CERA in 2012 (*Elton*). Another initiative saw the C&CT and the CCC joining forces to draft the *Christchurch Sports Tourism Events Plan* released in May 2013 (CCC & C&CT, 2013). However, the plan fell short following the establishment of the Sports Tourism and Event Group with other relevant stakeholders (*Elton* and *Joseph*). More recently, the CDC, CCC, C&CT, Vbase and the Christchurch International Airport began working on a new visitor strategy replacing the *Greater Christchurch Visitor Recovery Plan*. *Ted* stressed the importance of this collaboration in the effort of re-branding Christchurch as a vibrant city following the earthquakes. Eventually, this initiative paved the way for the establishment of a new city marketing body (*Colin* and *Ted*).

Overall, participants acknowledged that the earthquakes of 2010 and 2011exposed the fragmented governance of Christchurch as a tourist destination (*Elton*). Moreover, the recovery authorities heightened the divisions between tourism-relevant stakeholders, with the C&CT representatives 'always feeling a little bit disengaged' (*Adrian*). At the time of fieldwork, tourism-relevant stakeholders expressed frustration and disappointment with the stage of recovery (*Colin, Elton* and *Jeb*).

Discussion

The findings of this study are similar to those addressed in the literature on stakeholder analysis in tourism policy and planning studies. First, the complexity of the post-earthquake metagovernance of Christchurch with relevance to tourism underpins previous research conducted in urban tourism destinations (Stevenson *et al.*, 2008). Second, the influence of external tourism-relevant stakeholders like CERA and the CCDU resembles the findings from Rügen, former East Germany, where external stakeholders 'became powerful local economic actors and gained influence in the island's politics' (Bramwell & Meyer, 2007: 785) following the unification. Third, the use of mechanisms such as the earthquake legislation in Christchurch underpins findings from Australia, where 'special legislation and closed-door negotiations' (Dredge & Whitford, 2011: 494) shaped the governance for tourism and events. Finally, the tensions among tourism-relevant stakeholders which emerged from this study underpins previous research undertaken in Sapa, Vietnam (Truong *et al.*, 2014).

The study suggests that the values and beliefs of the recovery authorities were at odds with those of tourism-relevant stakeholders in post-earthquake Christchurch. Tourism-relevant stakeholders and organizations had relatively weak policymaking and advisory powers. Moreover, the unawareness of sound tourism development strategies within CERA and the CCDU greatly contributed to metagovernance failures. These findings underpin research conducted in Cyprus, where government authorities supported the industry interests with the implementation of large-scale tourism developments (Farmaki *et al.*, 2015; Stevenson *et al.*, 2008). Similarly, in Christchurch, the Convention Centre Precinct project openly benefited local real estate mogul, Philip Carter.

The metagovernance structure of Christchurch following the earthquakes hindered the recovery of the city as a tourist destination. A similar finding was found in post-tsunami Thailand, where 'increased community disillusionment in governance processes, and left communities with no effective representative mechanism at the local level' (Calgaro *et al.*, 2013: 14). Tourism-relevant stakeholders encountered difficulties in engaging with the recovery authorities that were similar to those of grassroot and community organizations in Christchurch (Amore *et al.*, 2017; Cretney, 2017). The findings from Christchurch suggest that not all tourism and hospitality stakeholders benefit from the changes in the metagovernance structure. However, it partly agrees with those works stressing the influence of big developers in the redevelopment of spaces following a natural disaster (Gotham & Greenberg, 2014; Gunewardena, 2008; Stonich, 2008).

This study suggests that failure to implement successful tourism strategies in times of crisis and uncertainty 'is largely the result of incompetent administrations' (Farmaki *et al.*, 2015: 186) in which 'diverse interests, due to the multiple stakeholders involved in tourism, create a complicated setting where power struggles over authority, resource utilisation and decision-making dominate' (Farmaki *et al.*, 2015: 187). The lack of efficient planning and coordination among tourism-relevant stakeholders acted as a barrier to sound tourism strategies. Particularly in Christchurch, the hierarchical metagovernance led by CERA, the CCDU and the Minister for the Canterbury Earthquake Recovery put tourism-related stakeholders in a situation in which a narrow group of politicians and bureaucrats made detrimental decisions on the future of Christchurch as a tourist city behind closed doors.

This study contradicts Orchiston and Higham (2016) with regards to the development of broader industry networks as result of the earthquakes. The earthquake legislation empowered CERA and the CCDU to pursue an anchor-project-driven recovery strategy without consulting tourism-relevant stakeholders. Arguably, the different approach in the selection of participants and the time of the fieldwork can explain the different findings of the two studies. On the one hand, this study recruited tourism-relevant stakeholders in Christchurch between 2013 and 2015. On the other hand, Orchiston and Higham (2016) recruited participants from national and local tourism-dedicated organizations like the C&CT, Tourism New Zealand and TIANZ between 2011 and 2012. Nonetheless, this work argues that metagovernance failures in post-earthquake Christchurch much contributed to the crisis of tourism in the city.

Conclusions

This study looked at the wants, needs and degree of collaboration among tourism-relevant stakeholders in post-earthquake Christchurch. It acknowledged the fragmented nature of the tourism sector and looked at governance and metagovernance failures relevant to tourism over a period in which the rhetoric of redevelopment as opportunity legitimizes extraordinary changes in the governance structure of destinations.

This study sheds light on the changes in the governance following a major natural disaster. Particularly with relevance to tourism-related stakeholders, the few works available suggested that post-disaster metagovernance paved the way for more efficient and collaborative partnerships. This study, instead, shows how the earthquakes of 2010 and 2011 negatively affected the degree of collaboration among tourism-relevant stakeholders in Christchurch. Moreover, this study uses social theory and planning theory to move beyond 'old ways of knowing' (Dredge & Jenkins, 2011: 5) in current tourism research and provides an enhanced theoretical basis to understand often-implicit assumptions regarding the role of the state, businesses, civic society and individuals in urban destination governance and planning (Amore & Hall, 2016b; Bramwell, 2011).

As stressed in other publications, there is need for more research on the key episodes of governance following the disestablishment of CERA and the CCDU in April 2016 (Amore & Hall, 2016a, 2017). The most controversial anchor projects are currently under way or soon to break ground (Otakaro Limited, 2017, 2018). Further research is necessary to extend the period of analysis to 2022 and beyond. This would enable scholars to fully appraise the process of recovery, from the drafting of the recovery strategy down to the implementation of the plans relevant for the redevelopment of Christchurch as a tourist destination.

References

Amore, A. and Hall, C.M. (2016a) 'Regeneration is the focus now': Anchor projects and delivering a new CBD for Christchurch. In C.M. Hall, S. Malinen, R. Vosslamber and R. Wordsworth (eds) *Business and Post Disaster Management: Business, Organisational and Consumer Resilience and the Christchurch Earthquakes* (pp. 181–100). Abingdon: Routledge.

Amore, A. and Hall, C.M. (2016b) From governance to meta-governance in tourism?: Re-incorporating politics, interests and values in the analysis of tourism governance. *Tourism Recreation Research* 41 (2), 109–122.

Amore, A. and Hall, C.M. (2017) Sports and event-led regeneration strategies in post-earthquake Christchurch. In N. Wise and J. Harris (eds) *Regeneration, Events, Sport and Tourism* (pp. 100–118). Abingdon: Routledge.

Amore, A., Hall, C.M. and Jenkins, J.M. (2017) They never said 'Come here and let's talk about it': Exclusion and non-decision making in the rebuild of Christchurch, New Zealand. *Local Economy* 32 (7), 617–639.

Arts Voice (2011) *Submission to CCC Draft Central City Plan by Arts Voice*. Christchurch: Arts Voice.

Becken, S. (2013) Developing a framework for assessing resilience of tourism sub-systems to climatic factors. *Annals of Tourism Research* 43, 506–528.

Biggs, D., Hall, C.M. and Stoeckl, N. (2012) The resilience of formal and informal tourism enterprises to disasters: Reef tourism in Phuket, Thailand. *Journal of Sustainable Tourism* 20 (5), 645–665.

Bramwell, B. (2011) Governance, the state and sustainable tourism: A political economy approach. *Journal of Sustainable Tourism* 19 (4–5), 459–477.

Bramwell, B. and Meyer, D. (2007) Power and tourism policy relations in transition. *Annals of Tourism Research* 34 (3), 766–788.

Calgaro, E., Dominey-Howes, D. and Lloyd, K. (2013) Application of the destination sustainability framework to explore the drivers of vulnerability and resilience in Thailand following the 2004 Indian Ocean tsunami. *Journal of Sustainable Tourism* 22 (3), 1–23.

Canterbury Development Corporation (CDC) (2014) *Christchurch Visitor Strategy. Background Paper*. Christchurch: CDC.

Carlsen, J. and Liburd, J. (2008) Developing a research agenda for tourism crisis management, market recovery and communication. *Journal of Travel & Tourism Marketing* 23 (2–4), 265–276.

Christchurch & Canterbury Tourism (C&CT) and Tourism Industry Association of New Zealand (TIANZ) (2011) *Joint Submission to the Christchurch City Council on the Draft Central City Plan for Christchurch Released August 2011*. Christchurch: C&CT.

Christchurch & Canterbury Tourism (C&CT) (2012a) *Greater Christchurch Visitor Recovery Plan*. Christchurch: C&CT.
Christchurch & Canterbury Tourism (C&CT) (2012b) *Christchurch and Canterbury Tourism. Annual Report for the Year Ending 30 June 2012*. Christchurch: C&CT.
Christchurch & Canterbury Tourism (C&CT) (2013) *Christchurch and Canterbury Tourism Annual Report for the Year Ending June 2013*. Christchurch: C&CT.
Christchurch & Canterbury Tourism (C&CT) (2015) *Eden Project. A Proposed Destination Tourism Project for the Recovery of Christchurch and the Benefit of New Zealand*. Christchurch: C&CT.
Christchurch Central Development Unit (CCDU) (2012a) *Christchurch Central Recovery Plan. Te Mahere 'Maraka Ōtautahi*. Christchurch: CCDU.
Christchurch Central Development Unit (CCDU) (2012b) *Christchurch Central Business District Blueprint - Implementing Anchor Projects*. Christchurch: CCDU.
Christchurch Central Development Unit (CCDU) (2013) FAQ – Roles. *CCDU*, Christchurch [online]. See http://ccdu.govt.nz/faq/roles.
Christchurch City Council (CCC) and Christchurch & Canterbury Tourism (C&CT) (2013) *Christchurch Sports Tourism Events Plan*. Christchurch: C&CT.
Christchurch City Council (CCC) (2011a) *Central City Plan. Draft Central City Recovery Plan for Ministerial Approval - December 2011*. Christchurch: CCC.
Christchurch City Council (CCC) (2011b) *Draft Central City Plan, August 2011. Technical Appendices*. Christchurch: CCC.
Christchurch City Council (CCC) (2017) *Mayor Welcomes Decision to Reinstate Cathedral (9 September)* [online]. See https://ccc.govt.nz/news-and-events/newsline/show/2001.
Convention & Incentive New Zealand (CINZ) (2014) *Convention Centre Good News for Tourism*. Auckland:.CINZ.
Cretney, R.M. (2017) Towards a critical geography of disaster recovery politics: Perspectives on crisis and hope. *Geography Compass* 11 (1), e12302. doi:10.1111/gec3.12302.
Cubrinovski, M. and McCahon, I. (2012) *CBD Foundation Damage. Short Term Recovery Project 7*. Christchurch: University of Canterbury.
Department of Prime Minister and the Cabinet (DPMC) (2012) *Cabinet Minute (12) 26/8. Christchurch Central Recovery Plan. Final Decisions by Group of Ministers with Power to Act. Minute of Decision*. Wellington: DPMC.
Dredge, D. and Jenkins, J.M. (2011) New spaces of tourism planning and policy. In D. Dredge and J.M. Jenkins (eds) *Stories of Practice: Tourism Policy and Planning* (pp. 1–11). Burlington, VT: Ashgate.
Dredge, D. and Whitford, M. (2011) Event tourism governance and the public sphere. *Journal of Sustainable Tourism* 19 (4–5), 479–499.
Evans, N. and Elphick, S. (2005) Models of crisis management: An evaluation of their value for strategic planning in the international travel industry. *International Journal of Tourism Research* 7 (3), 135–150.
Farmaki, A., Altinay, L., Botterill, D. and Hilke, S. (2015) Politics and sustainable tourism: The case of Cyprus. *Tourism Management* 47, 178–190.
Faulkner, B. and Vikulov, S. (2001) Katherine, washed out one day, back on track the next: A post-mortem of a tourism disaster. *Tourism Management* 22 (4), 331–344.
Gotham, K.F. and Greenberg, M. (2014) *Crisis Cities: Disaster and Redevelopment in New York and New Orleans*. Oxford: Oxford University Press.
Gunewardena, N. (2008) Peddling paradise, rebuilding serendib. The 100-meter refugees versus the toruism industry in post-tsunami Sri Lanka. In N. Gunewardena, M. Schuller and A. de Waal (eds) *Capitalizing on Catastrophe: Neoliberal Strategies in Disaster Reconstruction* (pp. 69–92). Lanham, MD: AltaMira Press.
Gurtner, Y.K. (2007) Phuket: Tsunami and tourism: A preliminary investigation. In E. Laws, B. Prideaux and K. Chon (eds) *Crisis Management in Tourism* (pp. 217–235). Wallingford: CABI.

Hall, C.M. (2008) *Tourism Planning: Policies, Processes and Relationships* (2nd edn). Harlow: Prentice Hall.
Hall, C.M. (2013) Financial crises in tourism and beyond: Connecting economic, resource and environmental securities. In G. Visser and S. Ferreira (eds) *Tourism and Crisis* (pp. 12–34). Abingdon: Routledge.
Hall, C.M., Malinen, S., Vosslamber, R. and Wordsworth, R. (eds) (2016) *Business and Post-Disaster Management: Business, Organisational and Consumer Resilience and the Christchurch Earthquakes*. Abingdon: Routledge.
Healey, P., de Magalhaes, C., Madanipour, A. and Pendlebury, J. (2003) Place, identity and local politics: Analysing intitiatives in deliberative governance. In M.A. Hajer and H. Wagenaar (eds) *Deliberative Policy Analysis: Understanding Governance in the Network Society* (pp. 60–87). Cambridge: Cambridge University Press.
Hunter, T. (2015) Convention centre already attracting business. *The Press*, 8 June. [online]. See https://www.stuff.co.nz/the-press/opinion/69119308/convention-centre-already-attracting-business.
Jamal, T. and Getz, D. (1999) Community roundtables for tourism-related conflicts: The dialectics of consensus and process structures. *Journal of Sustainable Tourism* 7 (3–4), 290–313.
Jessop, B. (2011) 'Metagovernance'. In M. Bevir (ed.) *The Sage Handbook of Governance* (pp. 106–123). London: Sage.
Johnson, L.A. and Mamula-Seadon, L. (2014) Transforming governance: How national policies and organizations for managing disaster recovery evolved following the 4 September 2010 and 22 February 2011 Canterbury earthquakes. *Earthquake Spectra* 30 (1), 577–605.
Johnson, L.A. and Olshansky, R.B. (2010) *Clear as Mud: Planning for the Rebuilding of New Orleans*. Chicago, IL: American Planning Association.
Larsen, R.K., Calgaro, E. and Thomalla, F. (2011) Governing resilience building in Thailand's tourism-dependent coastal communities: Conceptualising stakeholder agency in social–ecological systems. *Global Environmental Change* 21 (2), 481–491.
Lew, A.A. (2014) Scale, change and resilience in community tourism planning. *Tourism Geographies* 16 (1), 14–22.
McCrone, J. (2014) Great plan, but where are the investors? In B. Bennett, J. Dann, E. Johnson and R. Reynolds (eds) *Once in a Lifetime. City-Building After Disaster in Christchurch* (pp. 101–108). Christchurch: Freerange Press.
McDonald, L. (2016) Crown goes it alone on Christchurch Convention Centre after breakdown. *The Press*, 29 June [online]. See http://www.stuff.co.nz/business/81581654/crown-to-go-it-alone-on-christchurch-convention-centre.
McKercher, B. (1999) A chaos approach to tourism. *Tourism Management* 20 (4), 425–434.
New Zealand Ministry of Business, Innovation and Employment (NZMBIE). (2010) *Spend by Visitors in Christchurch for the Year to March 2010*. Wellington: NZMBIE.
New Zealand Ministry of Business, Innovation and Employment (NZMBIE). (2015) *Canterbury Tourism Partnership*. Wellington: NZMBIE.
New Zealand Treasury (NZT). (2013) *2013 Budget Speech*. Wellington: NZT.
Olshansky, R.B., Hopkins, L.D. and Johnson, L.A. (2012) Disaster and recovery: Processes compressed in time. *Natural Hazards Review* 13 (3), 173–178.
Orchiston, C. (2012) Seismic risk scenario planning and sustainable tourism management: Christchurch and the Alpine Fault zone, South Island, New Zealand. *Journal of Sustainable Tourism* 20 (1), 59–79.
Orchiston, C. and Higham, J.E.S. (2016) Knowledge management and tourism recovery (de)marketing: The Christchurch earthquakes 2010–2011. *Current Issues in Tourism* 19 (1), 64–84.
Orchiston, C., Vargo, J. and Seville, E. (2012) *Outcomes of the Canterbury Earthquake Sequence for Tourism Businesses*. Christchurch: Resilient Organisations.
Otakaro Limited (2017) Christchurch Convention Centre. *Otakaro Limited* [online]. See at https://www.otakaroltd.co.nz/anchor-projects/convention-centre/Timeline.

Otakaro Limited (2018) Multi-use Area. *Otakaro Limited* [online]. See https://www.otakaroltd.co.nz/anchor-projects/stadium/.

Reserve Bank of New Zealand (RBNZ) (2016) Bulletin Vol. 79, no. 3 – February 2016. *RBNZ*, Wellington [online]. See http://www.rbnz.govt.nz/-/media/ReserveBank/Files/Publications/Bulletins/2016/2016feb79-3.pdf.

Ritchie, B.W., Mair, J. and Walters, G. (2014) Tourism crises and disasters: Moving the research agenda forward. In A.A. Lew, C.M. Hall and A.M. Williams (eds) *A Companion to Tourism* (pp. 611–622). Oxford: Blackwell.

Scott, N., Laws, E. and Prideaux, B. (2008) Tourism crises and marketing recovery strategies. *Journal of Travel & Tourism Marketing* 19 (2–3), 149–158.

Stevenson, N., Airey, D. and Miller, G. (2008) Tourism policy making: The policymakers' perspectives. *Annals of Tourism Research* 35 (3), 732–750.

Stonich, S. (2008) International tourism and disaster capitalism. The case of Hurricane Mitch in Honduras. In N. Gunewardena, M. Schuller and A. de Waal (eds) *Capitalizing on Catastrophe: Neoliberal Strategies in Disaster Reconstruction* (pp. 47–68). Lanham, MD: AltaMira Press.

Treuren, G. and Lane, D. (2003) The tourism planning process in the context of organised interests, industry structure, state capacity, accumulation and sustainability. *Current Issues in Tourism* 6 (1), 1–22.

Truong, V.D., Hall, C.M. and Garry, T. (2014) Tourism and poverty alleviation: Perceptions and experiences of poor people in Sapa, Vietnam. *Journal of Sustainable Tourism* 22 (7), 1071–1089.

Wallace, S. and Simmons, D. (2012) *State of the Tourism Sector 2012*. Lincoln: Lincoln University.

Yuksel, F., Yuksel, A. and Bramwell, B. (1999) Stakeholder interviews and tourism planning at Pamukkale, Turkey. *Tourism Management* 20 (3), 351–60.

6 Bringing Relief to a Natural Disaster Zone Through 'Being a Tourist': The Case of the 2010 Earthquake in Haiti

Nigel D. Morpeth

Introduction

The overall purpose of this chapter is to review the ethics of bringing humanitarian aid and relief to a natural disaster zone through the intervention of travel companies and tourists in the immediate aftermath of an earthquake. The key chapter themes are in line with the overarching theme of the response of tourists, tourism organisations and tourism systems to disasters. Specifically considering how the concept of a disaster tourist might co-exist ethically in the same spatial territory as populations made the most vulnerable and at peril by the impacts of natural disasters, made more excessive by the lack of preparedness and community resilience. Within these considerations is a brief review of the early history of attempts to articulate, at least, if not to implement these voluntary inspired initiatives (Kelman & Dodds, 2009).

Whilst there are expected interventions of post-disaster relief traditionally provided by NGOs, relief funds and a range of actors within the international community, the intervention of tourist organisations and tourists within the immediate post disaster zones raises questions of the ethics of 'being a tourist' in the midst of a disaster zone. Increasingly through media revelations, it also raises important questions of both the motives of NGOs and relief agencies, who traditionally had key roles in humanitarian relief and now the motives of some of their employees have drawn them into an ethical debate as to an abuse of fundamental human rights and dignity. Coupled with this are the motives of travel companies, taking tourists to private beaches in a disaster zone and in the case of Haiti, a location that might suffer further aftershocks after the earthquake of 2010 but starkly to a location close to mass fatalities. In order for this performance to take place, there is a suspension of reality best described by tourists visiting a non-place (Auge, 1995). This ethical debate is made more complex by oligarchical travel organisations, such as cruise companies who are able to dominate the economies of developing nations through the rehearsed choreography of a transnational capitalist class (Sprague-Silgado, 2017).

Therefore, this chapter will explore the ethical debate highlighting the potential benefits and dis-benefits of the intervention of a travel company and tourists within the disaster zone of an earthquake in Haiti in 2010. It will also determine how a lack of preparedness for natural disasters in developing countries, such as Haiti, might intensify the human cost of natural disasters. It will evaluate the capacity of the Haitian government to improve its capacity for emergency responses to mitigate to future impacts natural disasters and whether Khasalamwa's (2009) 'concept of build back better' has emerged in Haiti.

Sociological and anthropological insights into the social construction of disasters

Faulkner (2001) was one of the first tourism academics to highlight that while tourist destinations might experience either person-made or natural disasters, few destinations had established preparedness for these eventualities through disaster management plans. Faulkner emphasised that it was possible to apply insights from wider disaster management literature, to better inform the tourist industry in planning for better preparedness for disasters. In doing so he implored the need for tourism disaster management plans, which would allow the tourism industry and destinations in which tourism occurred, to be more resilient.

In 2007 United Nations Environment Programme (UNEP) (2007: 1) highlighted both the range and magnitude of natural disasters stating that 'Every year more than 20 million people are affected by droughts, floods and cyclones, tsunamis, earthquakes, wildfires and other disasters associated with natural hazards. Growing populations, environmental degradation and global warming and are making the impacts worse'. UNEP (2007) went on to say that preparedness for these eventualities should be part of the integrated management plans for any destination, whilst also cognisant that the structuring of societal factors, will vary according to development patterns across the globe, and might make this a hollow aspiration. Crucial to unfolding insights within this chapter, are the observations of Cohen (2007: 21) who identified a new paradigm in the study of disaster, emphasising that disasters 'rather than caused by factors exogenous to the social system are at least partially socially constructed or produced'.

Therefore, in relation to the tourism industry, he felt that 'it' was culpable in disregarding disaster hazards and that social and anthropological perspectives enable insights to be developed, in which disasters are contingent, increasingly, on human activity on the environment, both globally and locally. An unvirtuous circle of local environmental degradation leading to a grand-narrative of global warming. This in his view both increases the likely incidence of 'natural' disasters and serves to intensify their effects. While Cohen's insights were focused on natural disasters in South-East Asia, the tenets of his arguments, are germane to the country of Haiti, positioned within the Caribbean, on the island of Hispaniola (shared with the Dominican Republic) which Bilham (2010) identified endured the most injurious earthquake in 200 years, on 12 January 2012. As Bilham (2010: 878) observed 'With an official death toll of 230,000 and thousands still buried beneath collapsed structures, the Haiti earthquake of 12 January was more than twice as lethal as any of the previous magnitude 7.0 event'.

Extending Cohen's observation that the intensity of the impacts of natural disasters are in part contingent on the societal factors that are social constructed, Wamai and Larkin (2011: 56) highlight how 'Haiti's history is marred by neo-colonialism, structural violence, dictatorial politics, and severe natural disasters. These social political and geo-ecological factors have played a strong role in shaping the country's past and current experiences in health and development' and emphasise that Haiti is defined by complex social and economic factors, which underpin the characteristics of a population of 9 million people, defined by 'Haiti [as being] the poorest country in the Western hemisphere' (Wamai & Larkin, 2011: 56).

Haiti contrasts with other Caribbean islands in terms of its recent history and the way in which the combination of natural geographical factors and the political and socioeconomic conditions have led its increased susceptibility to mega-crises and disasters. Mendel-Forman and White (2011) highlighted that Haiti is part of the island of Quisqueya-Hispaniola which also includes the state of the Dominican Republic. Whilst there is this geographical proximity between the two nations, increasingly there is a realisation of the susceptibility for the Dominican Republic of sharing the Enriquillo-Plantain Garden Fault with Haiti. With its shifting geography, this fault line increasingly does not privilege one country over the other in terms of the susceptibility to the devastating impacts of natural disasters. Mendel-Forman and White (2011: 11) noted that 'Haiti's annual GDP before the earthquake never rose above US$6 billion; the Dominican Republic's is well over US$50 billion. Haiti places 115 out of 137 on the Millennium Development Global Progress Index; in contrast, the Dominican Republic ranks 34. A startling 80 percent of Haitians live in poverty compared to only 40 percent of Dominicans'. Mendel-Forman and White (2011) also observed that prior to the January 2010 earthquake these two states whilst sharing a 'porous border' and shared environmental ecosystem but equally a mistrust and a lack of cooperation between the two states. Therefore, in terms of preparedness for a natural disaster, Haiti was poorly prepared on a socioeconomic level because of the broad level of poverty of the indigenous population exacerbated by the debilitation and crushing of democratic aspirations under the violent Duvalier regimes which dominated everyday lives of Haitians for the latter part of the 20th century. A largely under-educated and rural workforce which Bilham (2010) identified as forming 96% of the population.

Being a Disaster Tourist

Kelmal and Dodds (2009: 272) stated that 'Many disasters not only involve tourists but also attract tourists, with the disaster situations and their commemorations leading to "disaster tourism"'. They defined disaster tourism as 'travel for predominantly recreational or leisure purposes to see areas affected by a disaster defined by UNISDR (2008)' (Kelmal & Dodds, 2009: 273). (UNISDR is the United Nations International Strategy for Disaster Reduction). They argued that this definition should exclude tourism (including people travelling in a professional capacity or as a volunteer) to destinations to engage in disaster mitigation, preparedness or prevention projects. However, in their view, it would encompass travel to historical disaster sites, as well to post-disaster sites related to both recovery and memoralisation (Kelmal & Dodds, 2009). They contrasted this definition of disaster tourism, with observations on definitions of dark or thanatourism, which they viewed more broadly to visitation to sites of death and atrocity, locations which do not inevitably correspond to disaster sites and particularly natural disasters. They provided a range of examples of how natural disasters have been the source of tourism in diverse settings, including tourists who went purposively to Mt. St. Helens (in Washington State, US) in 1980 and included 20 of the 57 fatalities who perished in the aftermath of the volcanic eruption(ibid). Additionally, they cite examples in both 2004 and 2005, of tourists travelling to beaches in South East Asia and the Indian Sub-Continent, to be voyeurs to witness the impacts of tsunamis. As well as citing examples of self-motivated travel, they also highlighted how the travel industry created orchestrated tours to disaster locations through US based storm-chasing tours and specifically to New Orleans in 2005, to witness the devastation to both property and people in the immediate aftermath of Hurricane Katrina. This last example particularly highlights the ethical conundrum created by companies that are willing to extract commercial gain out of the suffering of individuals and communities.

Kelman and Dodds (2009) argue that in broader terms, the tourism industry has attempted through certain initiatives, to embrace the complex challenge of creating and implementing a code of ethics for tourism. They cited the work of UNTWO (2001) in creating a *Global Code of Ethics* which in part, was a response to a consumer demand for moral ethics to underpin the tourism experience. This code of ethics was couched in general terms with forms of tourism, that should be in harmony with the 'attributes and traditions of host regions and communities'. Kelman and Dodds (2009) emphasised that ethical aspirations had been intensified through the proliferation of a responsible tourism industry. In part which might have responded as part of its industry aspirations, to the challenge of child exploitation through a global sex industry. The emergence of a responsible tourism industry might be underpinned and couched in the language of sustainability. Furthermore, they may be dominated by considerations of the *sustainability of the tourism industry and the perpetuation of its products*, as a first consideration, rather than how it might address ethical challenges. In commenting on these varying initiatives and industry trends, Kelman and Dodds (2009) noted that there was dominance and an absence of the emphasis on the ethics of tourism, on destinations impacted by disasters or more succinctly the concept of disaster tourism ethics.

Disasters and Tourism Ethics

Kelman and Dodds (2009: 279) in reflecting on the premise for more international aid and response to disasters, traditionally carried out by aid agencies and NGOs, is that the aim should be not 'to perpetuate the root causes that led to the need for aid in the first place'. This exposes two flaws in this maxim, in that as discussed already, the motivations and impact of aid agencies working on the aftermath of disasters are far from transparent and indeed it would appear with recent media revelations, honourable (a full discussion of this is beyond the scope of this chapter). Secondly, and most pertinent to the focus of this chapter, is Kelman and Dodd's (2009) observation that attempts at establishing codes of ethics for tourism, have not correspondingly addressed the ethics of tourism in disaster zones.

Kelman and Dodds (2009) were very adroit in balancing the multiplicity of considerations, that might underpin ethical considerations of the merits and de-merits of being a tourist in the aftermath of a disaster in a tourist destination. In terms of merits, it might be that tourist revenues could potentially enable the re-establishment of a tourism industry infrastructure. This is in turn, could help to establish increasing levels of visitation and by implication, enable benefits to be realised by host communities. Additionally, they argue that people who have experienced the impacts of disasters, both physical and psychological, might in theory be able to organise 'tours' which enable survivors of a disaster to cathartically 'tell their story'.

However, the converse of these potential advantages might in stark terms be that the: 'people enjoying or spending their leisure time (might be) exploring others' misery' (Kelman & Dodds, 2009: 277). They argue that, on a practical level, tourists might be 'competing' with locals for the use of basic resources and in the process denying locals access to food and water or may just get in the way of rescue efforts, not least 'body recovery' operations. This encroachment by tourists into places of suffering, in the aftermath of a disaster, might involve, insensitive recording of disaster sites, through photographs and filming, or might more overtly involve transgressions of trespass and theft. Kelman and Dodds (2009) emphasise that far from assisting directly in humanitarian aid or disaster relief, tourists might easily engage in 'gawking' and 'rubbernecking'. In doing so, Kelman and Dodds (2009) suggest that tourists might be exacerbating the psychological damage that people are already experiencing from the aftermath of a disaster.

In the absence of established tourist ethics, for disaster areas, Kelman and Dodds (2009) suggested that a code of ethics, should be underpinned by key considerations, which include, firstly safety consideration. In practical terms they suggest that disaster-affected people and responders should be prioritised in disasters and that this should also include rescue and body recovery operations. Secondly, in terms of risk imposition; individuals should not put other individuals at risk without consent. Thirdly with regards to 'authorities and rules;' there should be (within reason) adherence to the disaster-affected area's authorities' rules and regulations. Finally, they suggest that assistance offered in a disaster affected area, should consider the local context and should not just include disaster affected communities but potentially neighbouring communities not affected directly by the disaster (Kelman & Dodds, 2009: 282–283).

In highlighting these normative ethical expectations, they also recognised the challenges of implementing a code of ethics in a tourist industry, that is ham-struck by acts (or non-acts) of voluntary compliance. Therefore, in that respect they view the opportunity for 'convergence behaviour', between different individuals and organisations as difficult to achieve. Furthermore, in the absence of the establishment of a 'Disaster Tourism Operators Association' there was an emphasis on self-motivating tourists guided by advice from different sectors of the tourism industry, to observe ethical responses to disasters. *This raises the question of why tourists would wish to visit disaster zones as a form of leisure pursuit?*

Being a tourist in Haiti

The focus of this chapter is to raise this question in relation to Haiti. A tourist writing a review on Tripadvisor perceptively observed the nuances of a tourist destination 'owned' by the cruise company who transported tourists there. This highlights the complexity of how tourists might 'encounter' destinations perhaps devoid of local people characterised by privatised space, in which all but local workers can be excluded. Additionally, this tourist perceptively undergirds their observation with the seductive qualities of the destination's natural features. The tourist identified that 'We visited Labadee on a family cruise on the "Independence of the Seas". The main thing about Labadee to remember is. Its RCCL (Royal Caribbean Cruise Lines) property. They lease the land from Haiti, You aren't going to "Haiti", your basically going to an idealised RCCL version of it. That said, the beaches are nice and it is a nice spot' (www.tripadvisor.co.uk/showUser-Reviews-g147306-d-150250-v455108910-Labadee-Haiti.html). On the same TripAdvisor site listed, another tourist noted that 'A part of the island bought by Royal Caribbean was probably a nice place before they took over. Now it's a touristic place that gives you no feeling at all of being in Haiti'.

These views are redolent of the concept of 'anywhere land', where tourists are motivated to visit a composite or cloned type of destination, which does not need the moniker of a specific named destination. What the central consideration is, is that the attractiveness does not require an association with the cultural characteristics of a destination but which nevertheless fulfils a need to enjoy generic spatial elements, perhaps based on climate and outstanding natural geography. In the case of Labadee on Haiti, sun, sea and sand. This notion of 'anywhere lands' or 'cloned destinations' has parallels with Marc Auge's (1995) conceptualisation of the 'non-place', informed by his atomistic terms of 'de-localisation', 'homogenisation of culture' or 'world culture', which inform his concept of the non-place. These terms are part of the panoply of terms which inform the way in which we see the world in what he characterised as the 'super-modern', offering the view that 'The world of super-modernity does not exactly match the one in which we live, for we live in a world that we have not yet learned to look at. We have to relearn to think of space' (Auge, 1995: 35–36). These observations are helpful when extending analysis of the interaction of tourists visiting the port of Labadee on Haiti, particularly during the aftermath of the earthquake of January 2010.

Cruise tourism de-localisation and the creation of 'non-places'.

Sprague-Silgado (2017) extended themes expounded by Auge's (1995) conception of non-places, in his critical tourism trope of the impact of cruise tourism in the Carribbean region, not least within Haiti. In his analysis of the role of cruise tourists, he argues that cruise companies 'socially alienate them through the reality they experience and the ability to conceive of or determine the true character of what they temporarily interact with and inhabit' (Sprague-Silgado, 2017: 102). He argues that cruise companies as an exemplar of a model of the transnational capitalist class, 'micro-manage and control' both workers and tourists (Sprague-Silgado, 2017: 94).

The core of Sprague-Silgado's (2017) analysis, positions the oligopoly of the cruise tourism firms within the concept of a transnational capitalist class (TCC) and argues that this industry redefines corporate relations with the states that it interacts. This interaction is not motivated by national economic development but because of the demands made on states, are largely unaccountable This analysis has its foundations in the work of David Harvey and others, with Harvey's trope on the geopolitics of capitalism, which introduced the dynamics of the mobility of capital, not moving within national boundaries but within the privatised spaces of transnational corporations (Harvey, 1985).

Additionally, Sprague-Silgado (2017) argues that Cruise companies engage in racialised and gendered employment practices, which is characterised by low wages and non-unionisation made easier by operating under international 'flags of convenience'. The growth of cruise tourism within the Caribbean region, has seen the purchase and leasing of privatised beaches and holiday enclaves, keeping passengers in fenced and gated spaces. Furthermore, with control over bus tours, the taxi companies used and largely regulating the movement of passengers within privatised spaces, he identifies that 'passengers are channelled within the company controlled bubble' (Sprague-Silgado, 2017: 114). He characterises a globalised consumer culture which has seen tourists revenues in this region rising from $7 billion in 1996 to $28 billion in 2013 (Sprague-Silgado, 2017: 99) perhaps made easier by the cruise tourism industry being dominated by two companies; Royal Caribbean Cruises and Carnival Cruises, who between them own 70% of the cruise industry.

Sprague-Silgado (2017: 100) identifies in the context of the Caribbean region that 'only a very small strata of the Caribbean population gains long-term benefits from the business and local tax benefits are small and environmental damage significant'. Furthermore, he identifies that the cruise companies are able to circumnavigate regulatory regimes and 'sell "exotic" experiences to high consuming sectors while simultaneously exploiting workers and locals' (Sprague-Silgado, 2017: 102). Far from engendering the benefits to communities of national tourism development, he argues that companies such as Royal Caribbean Cruises, are able to undermine the 'national', in favour of developing transnational capital to the benefit of the growth of the Transnationalist capitalist companies.

Specifically, with regard to Haiti, Sprague-Silgado (2017: 114) identified it 'has served as a laboratory for the model of a privatised cruise port. During the final stages of the Duvalier regime in 1986, Royal Caribbean Cruises began cruises to Labadee a heavily guarded and fenced off private resort'. He added that this cruise operation, lease five privatised beaches and a forested peninsula from the Haitian state, which has 'run smoothly', except in emergency periods when the port has been shut down. The cruise tourists use these spaces for water-sports and purchase souvenirs from company owned outlets before returning to the ship at nightfall. In terms of employment, 300 workers are employed and Sprague-Silgado (2017: 108) identified that the company pay the Haitian state $6 per visitor while enjoying an annual revenue of $7.7 billion as a company. The cruise ships visiting Labadee have an occupancy of over 6000 cruise tourists (Sprague-Silgado, 2017: 115) and in his view maximise on board opportunities, for extracting profitable activities, which raises revenue streams for the company. This it is suggested would prevent the possibility of the benefits being maximised in countries such as Haiti and the neighbouring Dominican Republic state, which has the same model of 'privatised' ports. What is more, a fundamental element of Sprague-Silgado's (2017) trope of transnational capitalism, is that companies such as Royal Caribbean Cruises and Carnival Cruises are able to rely on ruling elites of countries, to offer their vital support in allowing the assimilation of national territory as central to the cruise ship experience. This is a trade-off for the diminution of the rights of workers.

These observations from Sprague-Silgado (2017) help to provide vital insights into the relationship of cruise companies and nation-states and provides clues to the dynamic of established visitor-host relationships, which might help to explain the underlying motivations of cruise companies during the immediate aftermath of a natural disaster.

The Haiti earthquake: The immediate aftermath

The journalist James Sturcke (2010) writing in *The Guardian* newspaper in the UK reports that 'Up to 100,000 people may have died in the Haiti earthquake, the Pan American Health Organisation, has said as the UN launches an appeal for $550m (£337m) in aid. More than 48 hours after the 7.0 magnitude earthquake struck, people clamoured for relatives missing in the rubble. Bodies lay all around the hilly city, with people covering their mouths and noses with cloth to block out the smell. Corpses were piled on pick-up trucks and delivered to the general hospital in Port-au-Prince, where the hospital director Guy La Roche estimated that there were 1500 bodies piled outside the morgue'. With such stark and shocking news of carnage and destruction there was an eerie juxtaposition, of demarcated areas of everyday life, where tourism and disaster were cheek-by-jowl and not viewed as natural bedfellows. Three days after the headline above, *The Guardian* announced the headline:

'No room in Haiti's cemeteries but cruise ships still find a berth'

The above headline appeared in *The Guardian* in the UK on 18 January, 2010 and signalled the plunging of tourism into a moral maze (Bilham, 2010). On first reading, it is difficult to create an 'ethical balance sheet', which offers 'light and shade', to what presumably would be viewed as an ethical 'no-brainer'. *How could people choose to be a tourist in a disaster zone and not offer immediate help to alleviate suffering but to act out 'being a tourist' with the hope that it would offer support to the injured and dying?* A further element to this moral conundrum is that in acting out 'being a tourist', people might not have given a thought to being in a disaster zone, seduced instead by a private beach paradise, a much anticipated stop on an expensive holiday itinerary. If one applies Sprague-Silgado's (2017: 114) description of cruise tourist's being 'micro-managed' by Royal Caribbean Royal Cruises and 'channelled within a controlled bubble'. It is easy to imagine that in that moment of arrival, Labadee became a 'non-place' or to paraphrase one of the tourists 'it had a nice beach and was a nice place'.

The Guardian journalist Robert Booth (2010a), writing on 18 January, six days after a devastating earthquake in Haiti, started his report with the detail that 'Sixty miles from Haiti's devastated earthquake zone, luxury liners dock at private beaches where passengers enjoy jet ski rides, parasailing and rum cocktails delivered to their hammocks'. The article reported on two cruise ships docking and tourists enjoying the joy of five private beaches and wooded peninsula and enjoying water-sports, a zip-line excursion and souvenir shopping under the watchful eyes of armed guards, direct agents of Royal Caribbean's Cruises transnational capitalist class project.

So what are the potential positives of going through with the planned visit to Labadee as part of the cruise ship itinerary?

Robert Booth (2010a) quoted the words of the then vice-president, John Weis: 'In our conversations with the UN special envoy of the government of Haiti, Leslie Voltaire, he notes that Haiti will benefit from the revenues that are generated with each call ... We also have tremendous opportunities to use our ships as transport vessels for relief supplies and personnel to Haiti. Simply put, we cannot abandon Haiti now that they need us now'. Booth (2010a) identified that relief supplies included 'Forty pallets of rice, beans, powdered milk, water and canned foods were delivered on Friday, and a further 80 were due today and 16 on two subsequent ships. When supplies arrive in Labadee, they are distributed by Food for the Poor, a long-time partner of Royal Caribbean in Haiti'. Additionally, he reports that Royal Caribbean had pledged to donate $1 million to the relief effort, and apparently part of this figure would help 200 Haitian crew members. This was in addition to the $55 million that the company had spent on modifications to the port in advance of the earthquake. John Weis (quoted in Booth, 2010a) highlighted that 'In the end, Labadee is critical to Haiti's recovery; hundreds of people rely on Labadee for their livelihood' adding that the decision to 'deliver a vacation experience so close to the epicentre of an earthquake' had been the result of detailed debate.

Primae facie it is possible to view these pronouncements, as positively adding to much needed aid for a disaster riven country, in which any offer of help in the immediate aftermath of the disaster might potentially be viewed as a positive. However, it was not clear whether the scenario of purely using the cruise ships to assist in aid, rather than to invite tourists to visit Labadee and symbolically present an image of 'business as usual'. Royal Caribbean Cruises, might have also argued that their core business was tourism and the images of tourists enjoying themselves in a post-disaster zone, are still undergirded by a strong desire to offer humanitarian assistance at a time of mega-crisis. Furthermore, as an employer of Haitian nationalities, it could also be argued that there was a responsibility of providing economic benefits to this group of people at a time of crisis. Booth quoted one passenger who perhaps best embodies the 'business as usual philosophy': 'I'll be there on Tuesday and I plan on enjoying my zip-line experience excursion as well as my time on the beach'.

It is important to note that later that year in July 2010, at a meeting of the United Nations Economic and Social Council (2010) heard from Leslie Voltaire, in a session to discuss the recovery phase after the earthquake and the potential for a transition from aid to development, that less than two per cent of the promised $10 billion pledged to aid recovery had not been received by the Haitian government with no explanation as to why this had occurred.

Not all tourists on board the 'Independence of the Seas' decided to disembark on the heavily guarded port of Labadee. Booth (2010a) quoted one of the 'sickened' passengers who stated that 'I just can't see myself sunning on the beach, playing in the water, eating barbecue, and enjoying a cocktail while (in Port-au-Prince) there are tens of thousands of dead people being piled up on the streets, with the survivors stunned and looking for food and water'. Booth (2010a) reported that another passenger who echoed the ethical considerations expounded by the first tourist: 'It was hard enough to sit and eat a picnic lunch at Labadee before the quake, knowing how many Haitians were starving' and another said 'I can't imagine having to choke down a burger there now'. Booth (2010a) also captured the practical considerations of safety aspects communicated by tourists 'Some booked on ships scheduled to stop at Labadee are afraid that desperate people might breach the resort's 12 foot high fences to get food and drink'.

A separate report which appeared in *The Guardian* on 19 January announced somewhat enigmatically 'Cruise company to donate sun-loungers to Haiti make shift hospital'. In this article journalist Robert Booth (2010b) quoted Royal Caribbean's vice-president John Weis as saying that 'We've also gone through the site giving up any extra lounge chairs, bedding and mattresses which are now being used in the make shift hospital'. This article, as with the article, on 18 January, once again questioned the sensitivity of tourists arriving at Labadee, particularly that the port was deemed to be unsuitable to be used by aid ships. The framework for analysis provided by Kelman and Dodds (2009) suggests that on safety grounds alone, the risk taking movement of dropping tourists into a disaster zone prone to aftershocks would alone be a compelling argument not to place tourists in a zone that had killed up to 250,000 people. Let alone the myriad of ethical aspects which question the motives of a transnational corporation who has invested substantial funds into a privatised port.

The ethics of NGO and international aid interventions

Strong (2016) charted the historical origins of International Voluntary Service (IVS) and, in a more contemporary context, the emergence of NGOs and their impact on developing nations, particularly during time of duress and disaster. Her work is prescient in revealing the contradictions of individuals imbued with humanitarian motivations of trying to challenge global poverty and inequality while part of NGO initiatives with vulnerable communities. This noble aspiration can nevertheless perversely create structural inequalities and a template for development dictated by the dominant role of NGOs. Speaking on BBC Radio 5 Live Breakfast Show, on 13 February 2018 the Haitian Ambassador to the United Kingdom observed that 'there are too many NGOs operating in Haiti'.

What has emerged in relation to this statement is international outrage and shock at the revelation that far from NGOs and aid agencies working to uphold and establish the human rights of communities in disaster zones, that international aid agencies have allegedly allowed their workers to sexually exploit vulnerable adults and children. These allegations included aid being given in return for sexual favours. In this respect the journalist Richard Vaughan (2018) writing under the headline in the newspaper on 13 February 'Oxfam's deputy chief quits over sex scandal', stated that 'The deputy chief executive of Oxfam quit yesterday over the handling of the scandal which has engulfed the aid sector, hours before the charity regulator announced it was opening an inquiry. Penny Lawrence became the first major scalp of the crises, saying that she took 'full responsibility for the behaviour of staff following allegations that Oxfam aid workers used prostitutes in Haiti and Chad.' (quoted in Vaughn, 2018) This article followed a feature in *The Guardian* on 10 February, which stated 'Oxfam pressed over claims aid workers in Haiti used prostitutes' and revealed that 'Among those that quit without disciplinary action was Oxfam's country director in Haiti, Roland van Hauwermeiren, who, according to the report (a confidential report seen by the Time newspaper in 2011) admitted using prostitutes at the villa rented for him by Oxfam' (Grierson, 2018).

In providing a broader perspective on these revelations, journalist Richard Dowden (2018) explored the basis of why aid agencies exist stating that the 'Aim of aid agencies should be their own abolition'. Perhaps with parallels to Sprague-Silgado's (2017) trope on the growth and all pervading reach of transnational capitalist class (TCC), Dowden (2018) highlighted that 'What started as a relief agency in areas of hunger at the end of the Second World War has grown into a vast industry – a leader in international development assistance that has become more and more corporate'. In direct reference to Oxfam he identified that 'Led by overpaid staff in United Nations organisations, they have become far removed from the people they are supposed to serve, except-apparently-when it comes to sex' (Dowden, 2018) Not only in his view had aid agencies become more corporate, that there was competition between aid agencies. He also observed that the Department for International Development had dispersed part of their budget to aid agencies 'making their organisations and their bosses more powerful' (Dowden, 2018). Whilst it is suggested that claims of human rights abuse of disaster zone victims were known before these recent revelations they serve to question who or if outside agencies should be involved in the re-development of nation-states in the aftermath of natural disasters. Dowden (2018) concluded that 'I am not against aid. It has been essential in places suffering from war and disease. But real development can only be done by the people who live there. It cannot be parachuted in by outsiders.'

The UNDP in 2000-05 chose six countries that were characterised as stated affected by conflicts, with Haiti chosen as one of these case study states (Faubert, 2006). One of the aims of this project was to try to create human security in which there would 'freedom from poverty' and 'freedom from want'. The UNDP recognised that the structural reasons for poverty within the Haiti was in part due to the legacy of the ruling elite of the Duvalier who had been able to maintain their grip on power through creating a climate of fear underpinned by state driven violence. The UNDP highlighted that 'The situation in Haiti is not a post-conflict situation but rather a protracted and violent 20 year long transition following the end of the predatory dictatorship of the Duvaliers' (Faubert, 2006). So how does a country like Haiti move from a cycle of dependency on aid and move to a phase of growth and development? (Faubert, 2006).

'Building Back Better'

Whilst Khasalamwa's (2009) paper was written in the context of the post-2004 tsunami in Sri Lanka, her work is germane to natural disaster zones, per se, who are engaged in post-disaster reconstruction. In broad terms she identifies that a post-crisis phase, should respond to very complex needs of indigenous populations. These complex needs according to Khasalamwa (2009) should focus on the long-term goal of recovery based on understanding and addressing the 'root causes' or structural causes of vulnerability. Kolbe *et al.* (2013) posed the question 'Is Tourism Haiti's Magic Bullet? An Empirical Treatment of Haiti's Tourism Potential?' It is revealing that their report starts with the observation that in terms of a continuing cycle of aid dependency 'Haiti features amongst the highest number of active relief and development agencies in the world. There are everywhere between 3,000 and 10,000 non-governmental organisations (NGOs) in the country earning in the sobriquet – Republic of NGOs' (Kolbe *et al.*, 2013: 3). There was an acknowledgement that tourism to Haiti was dominated by VFR tourism but that there was scope to develop Pro-Poor Tourism as a way of developing community-based tourism initiatives. This form of tourism development corresponds closest to Khasalamwa's (2009) concept of tourism being part of the wider development goal of 'build back better'. But dominant motivations for tourism were still based on aspects of voluntourism and the role of missionaries in the country, suggesting that aid-dependency rather than growth and development are major features of national recovery. They concluded that whilst post-2010 there was a lack of an infrastructure for tourism there was nevertheless a perception from tourists that it was a relatively safe place to visit. They also highlighted a contrast between over half of VFR tourists being perceived as working class and another '10% live below the poverty line' (Kolbe *et al.*, 2013: 20). This was contrasted with expensive hotels that had been built in Petion-Ville targeted at wealthy business and individual travellers able to pay up to $150 dollars a night. The report revealingly makes no reference to cruise tourism.

Conclusion

It is clear from the accounts of tourists visiting the port of Labadee in Haiti post-2010 Earthquake that there is a lack of awareness of the recent history of the island and the level of suffering and destruction that the island has had to endure and recover from. Mendel-Forman and White (2011: 6) highlighted that 'It is now widely accepted that good governance and strong national institutions are critical to the reduction of disaster risk. In the current economic climate, many low-and middle-income countries are struggling with the governance of disaster risk and/ in particular, with the management of prospective risk'. In this respect contingency planning and risk assessment which characterised the pre-2010 earthquake lack of preparedness which might have mitigated against the wider impacts of a natural disaster. The post-disaster phase of 2010 merged into the impacts of the 2014 floods which affected an indigenous population still largely displaced from 2010.

Developing countries such as Haiti and countries such as Iceland have something in common in terms of experiencing the impacts of natural disasters. These countries have very little else in common in terms of the contrasts in the preparedness, contingency planning and risk assessment. Iceland has a long history of volcanic eruptions which are addressed through a National Risk Assessment prepared by the Department of Civil Protection. The Eyjafallajokull volcano eruption of April 2010 had both community and environmental impacts within Iceland but also closed European airspace and has a wider impact globally in terms of secondary impacts of businesses that were not able to transport goods into Europe. While Iceland was able to absorb the impacts of this natural disaster within a national context they were able to go one stage further and use this experience of this natural disaster to create a tourist visitor centre which allowed visitors to gain a wider understanding of the way in which a disaster can become part of the experience society. Bilham (2010) highlighted in terms of the reasons for the extent of the devastation of the 2010 earthquake highlighting that the 'buildings were doomed during construction' and that without some form of UN inspectorate system of construction 'the construction of buildings were designed to kill occupants' (Bilham, 2010: 878).

At the heart of the approach of transnationalist capitalist firms' response to states, such as Haiti, is perhaps the way in which indigenous Haiti's are viewed by visitors which may have enabled tourists to visit Labadee beach with impunity in the immediate aftermath of the January 2010 earthquake.

Perhaps underpinning ethical responses to nations might be contingent in the realm of international relations and the 'outing' of nations beyond the hegemony of the G8 and the dominant superpowers. Former President Bush's characterisation of the so-called axis of evil provided a populist approach to international diplomacy and the isolation of nations being on the edge of the 'civilised world'. This 'axis of evil' produces pariah nations that have to operate outside the normative expectations as how a nation should engage with other nations. This implies that the 'normality' of what perhaps nation-states under democratic systems offer the citizen. In truth the ability of so-called superpower nations to identify and isolate nations that don't conform to the norms and values of that society continues to skew people's perception of countries. In this respect, Winton *et al.* (2018) writing in *The Guardian* on 11 January 2018 highlighted that 'Donald Trump has been branded a shocking and shameful racist after it was credibly reported he had described African nations as a well as Haiti and El Salvador as "Shitholes" and questioned why so many of their citizens had ever been permitted to enter America.' This chapter therefore concludes that an ethics for tourism in post-disaster destinations is underpinned not only by the conduct of NGOs and aid agencies but also on the response and humanity of international politicians whose views can inform the 'othering' of nations and their citizens. Nations already isolated and stuck in a cycle of aid dependency and consigned to the purgatory on 'non-place' status (Auge, 1995).

References

Auge, M. (1995) *Non-Places: Introduction to an Anthropology of Super-modernity.* London: Verso.
Bilham, R. (2010) Lessons from the Haiti earthquake. *Nature* 463, 878–879.
Booth, R. (2010a) Cruise company to donate sun-loungers. *The Guardian,* 19 January.
Booth, R. (2010b) No room in Haiti's cemeteries but cruise ships still find a berth. *The Guardian,* 18 January.

Cohen, E. (2007) Tsunami and flash floods-contrasting modes of tourism disasters in Thailand. *Tourism Recreation Research* 32 (1), 21–39.

Dowden, R. (2018) Aim of aid agencies should be their own abolition. *inews*, 13 February.

Faubert, C. (2006) Case Study Haiti. Evaluation of UNDP Assistance to Conflict-affected Countries. United Nations Development Programme Evaluation Office [online]. See https://www.oecd.org/countries/haiti/44826404.pdf.

Faulkner, B. (2001) Towards a framework for tourism disaster management. *Tourism Management* 22 (2), 135–147.

Grierson, J. (2018) Oxfam pressed over claims aid workers in Haiti used prostitutes. *The Guardian,* 10 February.

Harvey, D. (1985) The geopolitics of capitalism. In D. Gregory and J. Urry (eds) *Social Relations and Spatial Structures* (pp. 128–163). Basingstoke: Macmillan.

Kelmal, F. and Dodds, R. (2009) Developing a code of ethics for disaster tourism. *International Journal of Mass Emergencies and Disaster* 27 (3), 272–296.

Khasalamwa, S. (2009) Is 'build back' a response to vulnerability? Analysis of the post-tsunami humanitarian intentions in Shri Lanka. *Norwegian Journal of Geography* 63 (1), 73–88.

Kolbe, A.R., Brookes, K. and Muggah, R. (2013) Is tourism Haiti's magic bullet? An empirical treatment of Haiti's tourism potential. *Strategic Note 9* IGARAPE Institute, 1–23.

Mendel-Forman, J. and White, S. (2011) The Dominican response to the Haiti earthquake: A neighbour's journey. *Centre for Strategic and International Studies:* Washington [online]. See http://csis-prod.s3.amazonaws.com/s3fs-public/legacy_files/files/publication/111208_MendelForman_DominicanResponse_WEB.pdf (accessed February 2018).

Sprague-Silgado, J. (2017) The Caribbean cruise ship business and the emergence of a transnational capitalist class. *Journal of World Systems Research* 23 (1), 93–125.

Sturcke, J. (2010) Haiti earthquake: Up to 100,000 may have died. *The Guardian*, 15 January.

Strong, A. (2016) *Volunteerism or Voluntourism? A Case Study of NGO Motivations and Success in South Africa*. Honors Diploma. Murray State University.

Tripadvisor review [online]. See www.tripadvisor.co.uk/showUserReviews-g147306-d-150250-v455108910-Labadee-Haiti.html.

United Nations Economic and Social Council (2010) *Less Than 2 Per cent of Promised Reconstruction Aid for Quake-Devastated Haiti Delivered, Haitian Government Envoy Tells Economic and Social Council*. Meetings Coverage, ECOSOC/6441, 13 July [online]. See at. www.un.org/press/en/2010/ecsoc6441.doc.htm (accessed February 2018).

United Nations Environment Programme (UNEP) (2007) *Disaster Risk Reduction in Tourism Destinations: Disaster Reduction through Awareness, Preparedness and Prevention Mechanisms in Coastal Settlements in Asia – Demonstrations in Tourism Destinations*. Nairobi: UNEP.

United Nations International Strategy for Disaster Reduction (UNISDR) (2008) *Terminology: Basic Terms of Disaster Risk Reduction*. Geneva: UNISDR.

UNWTO (2001) *Global Code of Ethics for Tourism*. Madrid: UNTWO.

Vaughan, R. (2018) Oxfam's deputy quits over sex scandal. *inews*, 13 February.

Wamai, R.G. and Larkin, C. (2011) Health development experiences in Haiti: What can be learned from the past to find a way forward? *Japan Medical Association Journal* 54 (1), 56–67.

Winton, P., Burke, J. and Livesy, A. (2018) 'There's no other word but racist': Trump's global rebuke for 'shithole' remark. *The Guardian*, 11 January.

7 The Tourism Industry Response in Assisting Resident Evacuees after the 2016 Kumamoto Earthquakes

Atsuko Hashimoto and David J. Telfer

Introduction

In April of 2016, a series of earthquakes occurred in the central Kumamoto region of Japan on the island of Kyushu. Two of the earthquakes were of considerable force including a magnitude 6.5 quake on 14 April and a magnitude 7.3 on 16 April (see Figure 7.1). At least 49 people were killed and 3000 were injured. Approximately 4600 buildings in Kumamoto Prefecture were destroyed in the April earthquakes along with serious damage to significant buildings including the Kumamoto Castle and Aso Shrine. The earthquakes triggered numerous landslides around Minamiaso Village located in the western part of the Aso caldera. The largest landslide (300 m high, 130–200 m wide) damaged National Route 57 and the Hōhi line of the Japan Railway (Miyabuchi 2016). Aftershocks continued for months. The events of 16 April also triggered a small eruption of Mount Aso volcano, which erupted again six months later on 9 October 2016. With homes destroyed or deemed uninhabitable due to damage or ground instability, many area residents were left with nowhere to live. Over 44,000 people were evacuated (Real Estate Japan, 2016). Aso Farmland, a neighbouring agritourism and health-themed hot spring resort has 480 earthquake resistant, polyurethane dome-shaped guesthouses. While constructed on the somma volcano of Mount Aso, none of the guesthouses were damaged in the earthquakes. In the aftermath of the disaster, Aso Farmland became an evacuation shelter for families with elderly residents and children at the request of the Minamiaso Government. The tourist resort took in 600 area residents in

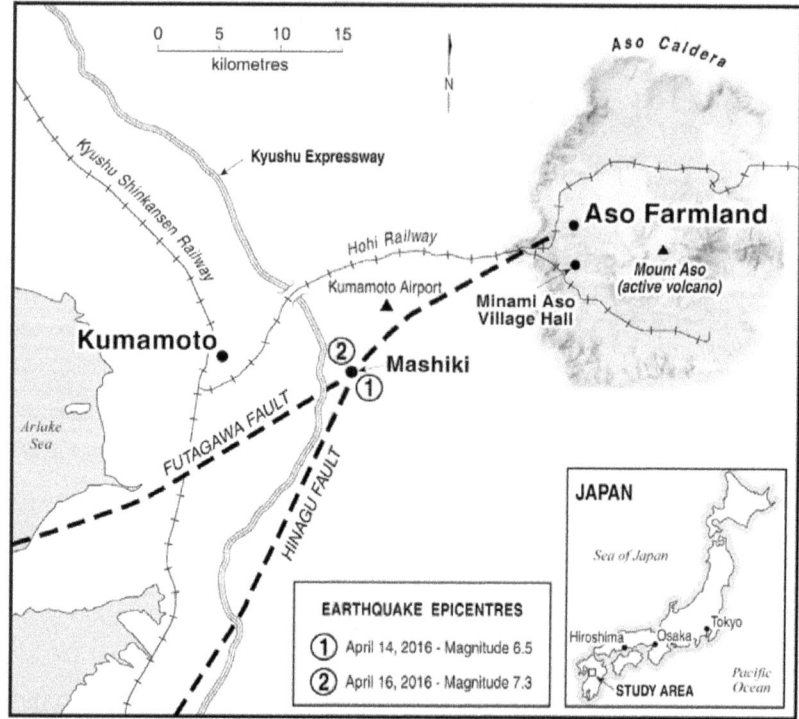

Figure 7.1 Kumamoto Earthquake epicentres and surrounding area (prepared for the authors by Loris Gasparotto, Cartographer, Brock University)

late May and was closed to tourists for several months (The Mainichi, 2016). Orchiston and Higham (2016) note the critical importance of knowledge management as well as effective inter-agency collaboration and communication in the immediate disaster response and tourism recovery. The collaboration during this crisis has led to new partnerships with tourism as Aso Farmland played a vital role in the disaster relief and recovery process. During the summer of 2017, the authors revisited the area and met with key informants to explore the impact of the earthquake, interagency collaboration, Aso Farmland's contribution to disaster relief and how the tourism industry can be linked into disaster planning.

This chapter begins by highlighting some of the key concepts of tourism crisis management focusing on collaboration and learning and then provides additional details on the Kumamoto mega-earthquake in April 2016. The tourism facilities of Aso Farmland are outlined, followed by an examination the stages the resort went through in being a disaster relief evacuation centre. The chapter concludes with lessons learned and the importance of collaboration in enabling the tourism industry to assist in disaster relief.

Collaboration and Learning in Tourism Crisis Knowledge Management

Schmidt and Berrell (2007: 68) define a crisis in terms of tourism as a 'low probability, high impact event that threatens the viability of tourism and its stakeholders, either directly or indirectly'. Parsons (1996) identifies three different types of crises: (1) immediate – little or no warning, (2) emerging – slower to develop and (3) sustained for weeks, month or years. In Kumamoto, there was the immediate crisis of the initial foreshock, as well as the sustained crisis the region is still facing. Several models of crisis management (e.g. Faulkner, 2001; Ritchie, 2004) focus on strategies under the categories of before, during and after the crisis. However, many tourism companies and destinations do not have crisis management plans (Hystad & Keller, 2008). Even within Japan, a country well known for disaster response, individual companies may not be fully prepared. Japan learned a great deal from the 2011 Tohoku earthquake and so the response in Kumamoto was very fast, yet the region still struggles with the duration of the crisis. There is increasing recognition of the importance of crisis knowledge management 'to improve destination recovery and resilience to future events, and in communicating to all relevant stakeholders during a crisis' (Orchiston & Higham, 2016: 67). Ghaderi *et al.* (2014: 627–628) indicate that there is a lack of 'investigations into how organisations that were involved in tourism crisis management have learned from their previous experiences'. They propose an integrative model of organizational learning for crisis management and put reflection at the centre. Similarly, collaborative approaches to emergency management are beginning to challenge command and control approaches (Morakabati *et al.*, 2017). Scott *et al.* (2007) present a conceptual model on the effects of a crisis on destinations that stresses the importance of learning systems. Before the crisis, firms have loose networks often operating independently; however, the effects of the crisis forges new stronger networks. Crises and disasters have been noted to have 'transformational connotations, with each such event having potential positive (e.g. stimulus to innovation, recognition of new markets, etc.) as well as negative outcomes' (Faulkner, 2001: 137). This stimulus for innovation can be linked to 'Building Back Better' a new concept referring to post-disaster recovery efforts where reconstruction is an opportunity to improve resilience in a community (Mannakkara *et al.*, 2018). This chapter explores some of the learning that has occurred and the new networks that Aso Farmland has developed. Becken and Hughley (2013) call for an increased role for tourism in wider emergency management structures as a way to enhance overall risk reduction. This case study illustrates how Aso Farmland has redefined how it collaborates with the local municipality and crisis response teams and at the same time has transformed its image and product.

Kumamoto Mega-Earthquake April 2016

At 21:26 on 14 April 2016, a magnitude 6 (M_w 6.2) earthquake hit Kumamoto city. The epicentre was 11 km below Mashiki-town, at the eastern end of Kumamoto city where two fault lines (Hinagu Fault and Futagawa Fault) meet. To the residents' dismay, this was only a foreshock. A bigger earthquake of M_w 7.1 (M_{jma} 7.3) hit Mashiki-town again at 1:25 on 16 April. This earthquake of M_w 5.4 was also recorded simultaneously in the neighbouring Prefecture of Oita. Table 7.1 includes a list of the earthquakes from 14 to 31 August 2016.

The initial foreshock (M_w 6+) on the 14th of April was felt over a large area including the entire Kumamoto City area, stretching to the southwest (to Uki city) and to the northeast (to Aso city) of the epicentre, Mashiki-town (M_w 7). There were a total of four large foreshocks ($M_w \geq 5.5$) on the same day. The epicentre for the main shock on 16 April, was again Mashiki-town (M_w 7.3). Minamiaso village, where Aso Farmland is located, also felt the shock of M_w 6+. This main

Table 7.1 Kumamoto Earthquake (earthquakes recorded stronger than M 5.0) between 14 April and 31 August 2016

Date	Time	Area (epicentre)	Depth of epicentre (km)	Scale of magnitude (M)	JMA seismic intensity (max)
14 April	21:26	Kumamoto	11	6.5	7
	22:07	Kumamoto	8	5.8	6–
	22:38	Kumamoto	11	5.0	5–
	23:43	Kumamoto	14	5.1	5–
15 April	00:03	Kumamoto	7	6.4	6+
	00:06	Kumamoto	11	5.0	5+
16 April	01:25	Kumamoto	12	7.3	7
	01:25	(Oita Central)	12	5.7	–
	01:30	Kumamoto		5.3	4
	01:44	Kumamoto	15	5.4	5–
	01:45	Kumamoto	11	5.9	6–
	03:03	Aso	7	5.9	5+
	03:55	Aso	11	5.8	6+
	07:11	(Oita Central)	6	5.4	5–
	09:48	Kumamoto	16	5.4	6–
	16:02	Kumamoto	12	5.4	5–
18 April	20:41	Aso	9	5.8	5+
19 April	17:52	Kumamoto	10	5.5	5+
	20:47	Kumamoto	11	5.0	5–
31 August	19:46	Kumamoto	n/a	5.2	5–

Source: Japan Meteorological Agency (2016a, 2017a).

shock was followed by four large aftershocks of $M_w \geq 5.0$ that occurred within 6 hours (Lin et al., 2016; Aoi, 2016; Japan Meteorological Agency, 2016a). A characteristic of this Kumamoto earthquake was long-period ground motion. The epicentre Mashiki-town recorded mainly short-period ground motions of 1–2 seconds, causing great damage to low-rise buildings; on the other hand to the east of Kumamoto city, just west of the Aso Caldera, Nishihara village recorded long-period ground motion of 2–5 seconds causing more serious damage to high-rise buildings (NHK (Japan Broadcasting Corporation), 2017; NHK Kabun, 2016; Japan Meteorological Agency, 2018a). The long-period ground motions in Minami Aso village and Nishihara village were categorized as level 4 (highest) in which a person 'is unable to keep standing; can move only by crawling; has no control over ground shake' (Japan Meteorological Agency, 2016b). This earthquake exhibited a stronger shock as it spread south from the epicentre (Takayama, 2017).

Other dangers were the numerous aftershocks (Table 7.2). According to the Japan Meteorological Agency (2017b), 556 aftershocks of $M_w \geq 3$ were recorded between 14 April 2016 and 30 June 2017. An aftershock of M_w 4 was recorded once on 08 September 2017; M_w 3 on 11 November 2017 and M_w 1 or 2 were recorded on 30 January 2018 at the time of writing this chapter (Japan Metrological Agency, 2018b). Having such small, but prolonged aftershocks, caused more damage to structures and landscapes already weakened by the stronger earthquakes.

With two M_w 7-level shocks and numerous aftershocks, damage to infrastructure was considerable. According to the Civil Engineering Department of the Kumamoto Prefectural Government, motorways (Kyushu Expressway and Minami-Kyushu Expressway) suffered surface damage in 350 areas (ruptures, sliding, twisting, etc.) including on 13 main

Table 7.2 Earthquake frequency by magnitudes (14 April 2016–30 June 2017)

Dates	Earthquake frequencies by magnitude						
	M7	M6	M5	M4	M3	M2	M1
2016/4/14-4/30	2	5	15	98	323	859	1722
5/1-5/31				8	43	134	344
6/1-6/30			1	4	14	51	147
7/1-7/31				1	8	19	85
8/1-8/31			1	2	3	28	77
9/1-12/31				4	15	52	145
2017/1/1-6/30				3	6	37	92
Total	2	5	17	120	412	1180	2612
Total ≥ M3					556		
Total ≥ M1							4348

Source: Created from Japan Meteorological Agency (2017b).

bridges and five smaller bridges, with one bridge section collapsing. The Kyushu Shinkansen line was quickly repaired and reopened within a week, but other railways were not so fortunate. The JR Hōhi line is still not fully operational as of March 2018 (JR Kyushu, 2018). The Minamaiaso Testudō line was partially operational by the end of July 2016; however, the section through Tateno will require over five years to repair. The main shock of the earthquake moved Saikaku mountain, causing a distortion of the rail tunnel by approximately 49 cm (Norimono News, 2017). Aso Kumamoto International Airport in Mashiki town suffered building damage, particularly to the domestic terminal. While all passenger flights were cancelled from 16 to 18 April 2016, disaster relief cargo was delivered to Kumamoto Airport. Domestic flights resumed on 19 April after quick repairs (International Civil Aviation Organization (ICAO), 2016), however, international airlines resumed services somewhat later. China Airlines resumed flights to Kaoshun on 03 June 2016 (Aso Kumamoto Airport, 2016); T'way Air and Air Seoul resumed in April 2017 and October 2017, respectively; and Hong Kong Express restarted in November 2017 (Tagami, 2017). The Ministry of Land, Infrastructure, Transport and Tourism (MLIT) (2018) decided to outsource operations at Kumamoto Airport to the private sector from 2020 in hopes of further accelerating the earthquake recovery.

Direct Impacts of Kumamoto Earthquakes on the Aso Area

The City of Aso includes volcanic mountains situated inside and outside the Aso Caldera. Minamiaso village, which is not part of City of Aso, is located inside the Aso Caldera, and occupies the southwest area of caldera. Since Aso City borders with Minamiaso village in the south, these areas are often combined and referred to as the Aso Area. The Kumamoto Earthquake caused a number of coseismic deformations:

> The coseismic surface ruptures were generally concentrated in a zone that can be divided into three segments (southwest, central, and northeast) ... The southwest and central segments consist mainly of distinct shear faults, left-stepping echelon cracks, and additional tracks constrained to a zone 3–50 m wide (average width, 5–10 m) ... the shear faults are dominated by right-lateral strike-slip displacement. Horizontal striations on slickensides were observed on shear fault planes; the striations are marked by parallel lineations with some grooves and steps in unconsolidated clay... (Lin et al., 2016)

Cracks on the roads, hillsides, rice paddies, fields and wooded areas showed the sides of the rupture slid towards opposing directions. Fieldwork on the coseismic displacements by Lin et al. (2016) identified new fault lines and more importantly, the Magma chamber at the bottom of Aso Caldera played a significant role in 'terminating the seismogenic rupture propagation through the volcano and northeastward beyond the caldera'.

Miyabuchi's (2016) analysis revealed over 100 landslides in and around Minamaiaso village. After the main shock, on the north-western to western caldera walls, several landslides occurred on slopes steeper than 25°. This part of the caldera wall was made from pre-Aso volcanic rocks, i.e. lava and pyroclastic material. Meanwhile the largest landslide (*ca*. 300 m high, 130–200 m wide) occurred on the western caldera wall, near Tateno, severing National Route 57, and the Japan Railway Hōhi line and demolishing the bridge over the Shirakawa River. Miyabuchi's (2016) study also looked at the similarity of landslides to those formed in 2012 induced by rainfall. The patterns of landslides were distinctly different: he concluded that the 2016 earthquake-induced landslides at the post-caldera central cones of Aso Volcano, were generated not only on steep slopes but also on slopes gentler than 10°. Despite the low elevation and milder slopes, the sliding masses travelled long distances (<600 m) evidently washed further by water in the river. Miyabuchi (2016) also suggested that some landslides turned into debris avalanches, travelling a few hundred metres.

The main shock on 16 April caused multiple landslides in the Aso area. The landslide in Tateno area, Minamiaso, buried National Route 57 and one driver was killed. The same landslide and land erosion destroyed the bridge leading to National Route 325, which was the main road to Minamiaso village and tourist areas. Across the bridge, Tokai University residence buildings were razed or severely damaged and deemed structurally unsafe. Nearly 700 university students and 300 villagers in the area were stranded in the University Sports Gym complex on 16 April (Yoshikawa, 2016). The scenic switchback railway route in Tateno was also cut off. Landslides buried some houses, and the main shocks destroyed the foundations of houses or parts of houses.

Casualties of the 16 April earthquake in the Aso Area (both the City of Aso and Minamiaso Village) mounted with: 44 dead, 35 seriously injured, 218 injured: 805 buildings were totally destroyed, 1714 buildings were half destroyed and 2665 buildings were partially destroyed (Tokyo Metropolitan Government, 2016). As the Futagawa Fault ends at Minamaiaso village, it suffered greater damage than the City of Aso. The actual death toll in Aso village was 27, seriously injured 29 and injured 120. On 17 May 2016, Nishi Nippon Newspaper reported on the Minamiaso evacuation. In the initial stages, 99 temporary housing units were under construction, and as of 17 May applications had begun immediately. As of October 2016, a total of 4303 temporary housing units were built in Kumamoto Prefecture, of which 401 units were in Minamiaso village and 101 units were in the City of Aso (International Research Institute of Disaster Science (IRIDes), 2016). The secondary evacuation locations were designated as Aso Farmland (approximately

1000 households) and Vacation Village Minamiaso (70 households), and applications also begun on 17 May. Another complication was the heavy rainfall resulting in a total of 4765 people (1994 households) in the Tateno area in Minamiaso receiving an Evacuation Order to 10 safer locations and additional 6606 people (2693 households) receiving an Evacuation Alert (Cabinet Office, 2016a).

The earthquake caused geological shifts and interfered with the flow of underground hot water to tourist hot springs. The hydrothermal reservoirs approximately 50 m below the surface were pushed horizontally (>1 m) by the earthquake shocks. This either collapsed or bent the boreholes for hot spring intakes. The movement caused existing hot springs to dry up or pressurized the opening up of a new hot spring wells or caused seepage on the mountainside (Tsuji et al., 2017). This led to not only the closure of some commercial accommodations and spa resorts, but also caused inconveniences for locals who rely on the natural hot springs for bathing. Initially the Minamiaso Tourism Center could only contact less than half of its members and later confirmed the circumstances of approximately 150. As of 16 April, Aso Farmland suspended its operations as a tourist attraction due to ruptured roads and damage to their facilities. By 18 May, the repairs at Aso Farmland were still ongoing and a reopening date was uncertain (Nishinippon Shinbun, 2016).

Aso Farmland, Kumamoto, Japan

Under normal circumstances, Aso Farmland is a unique experiential tourist destination in rural Kumamoto located within the caldera of Mount Aso, in Kuju National Park. Well-placed road networks and railway systems have made it easily accessible from both Kumamoto city (to the west) and Oita city (to the east). However, accessibility was greatly obstructed by the Kumamoto Earthquake in April 2016. In this section, the facilities and character of Aso Farmland will be outlined, as this is vital to understanding this establishment's significance as a disaster relief evacuation centre.

Aso Farmland as a company and a destination

Aso Farmland was opened in 1995 with the theme of 'people, nature and vitality'. It calls itself a 'Health Theme Park' with the main concept of 'experiencing heath/wellness [kenkō 健康] with fun'. This unique tourism destination offering experiential multi-sensorial activities centred on health and wellness tourism through hot springs, physical activity and culinary tourism. It is run by a private corporation and receives approximately 4 million tourists annually including both domestic and international visitors (day trippers and overnight guests).

Aso Farm Land Co. Ltd is responsible for operations and is one of seven members of Aso National Park Healthful Forest Group. Other members include: Healthful Forest Co. Ltd., Okashijo Kagahan Co. Ltd., Japan Dome House Co. Ltd., International Dome House Co. Ltd. and Aso Biotech Co., Ltd. The main food supplier to Aso Farmland is an independent agricultural organization named Aso Heath Farm Co. Ltd. and focuses on producing food products without harmful chemicals (ASO National Park Healthful Forest, 2013; Aso Farmland, 2014).

The resort offers accommodation, spas, restaurants using local food, a forest of vitality (physical exercise zone) and shops featuring local farm and health products. In a review of the literature Medina-Muñoz and Medina-Muñoz (2013) suggest that wellness tourism has become a fashionable tourism product for a number of reasons including greater concern about health issues for the middle and upper classes, the promotion of physical, mental and or social wellbeing, coping with work stress as well as the need to get away from everyday life routine. A subsector of health and wellness and a major attraction at Aso Farmland are the hot springs and spas. 'Bathing in hot and mineral springs and drinking mineral water is part of this holistic approach to health and wellness' (Erfurt-Cooper & Cooper, 2009: 8). The resort is focused on recuperation and preventive therapies and as Japan is highly urbanised, it provides an opportunity for city dwellers to experience the rural environment.

Accommodation facilities of Aso Farmland

Aso Farmland has 480 domed, earthquake proof accommodation units that can house up to 1800 people per night. The units in the village are scattered over 10 hectares (24.5 acres). The priority was to maintain the natural landscape and geophysical environment, which led to the creation of individual dome-shaped accommodation buildings within a hilly area with streams, bridges, winding paths and narrow staircases (see Plate 7.1). A shuttle bus circulates within the accommodation 'village' for guests, or they can stroll along the tree-lined roads on their way to restaurants, exercises, bathing facilities and shops.

Each domed accommodation unit represents a guest's home and are seven metres in diameter, accommodating six guests or more. The domes are durable, low cost, have high thermal insulation values thereby saving energy and are earthquake and gale resistant. The units are made of polystyrene that includes an antioxidant solution, reputed to have anti-aging benefits. A circular window at the top of the dome provides extra light and there are no corners or support poles in order to establish a peaceful setting. The dome shaped has an open-concept design to reduce stress, provide unusual acoustics and light and enhance the circulation of air to positively impact the level of relaxation, and quality of sleep and rest.

Plate 7.1 Aso Farmland accommodation poster at the entrance to Aso Farmland complex (photograph of Aso Farmland Poster taken by David Telfer, June 2017)

The accommodation area is divided into three zones including the Village Zone, based on the standard domes, the Royal Zone where the units have a small garden and gate, and the Private House Zone. These private houses are geared to long-staying guests, and each house can be split-level or a two-storey structure.

Local food and farm market

One of the central tenets of the health program is to use local agricultural products in their 11 restaurants, which not only promote healthy living but also represent opportunities for regional branding and local development (Hall & Mitchell, 2002; Telfer, 2015). Aso Farmland is a working farm and also has food processing facilities on site. Visitors can tour the 'vegetable and mushroom cultivation factory'. Organic food production in greenhouses includes mushrooms, strawberries, leaf lettuce and herbs. They have their own dairy and beef farm as well as a brewery. Aso Farmland also purchases agricultural products from local organic farmers including tomatoes, carrots, pumpkins, corn and peppers. Outside the restaurants are posters with information detailing local farmers' contributions to restaurant menus. The restaurants range from larger buffets like the 'World Kitchen' offering cuisines from around the world to 'Harvest Restaurant' highlighting local recipes. The 'Healthy Style Restaurant Volcan' focuses on 'locally grown, locally consumed' food ingredients, featuring famous Aso Red Beef, served in western style cuisine. One of the specialties of the farm is mushroom production and these are highlighted in '*Kinoko* House', a mushroom speciality restaurant. The 'Healthy Hot Pot' restaurant allows diners to cook their own food over specially designed grills that drain away unhealthy fat in meat. Other restaurants specialize in Japanese cuisine, Korean cuisine, pizza and Kumamoto style noodle shops. The resort is home to a farmer's market, an attraction that is of interest to day trippers who also have the opportunity to purchase Aso Farmland produce.

Onsen baths (hot springs) and heated saunas (earth power spas)

Japan is well know for hot springs (*onsen*) (Cooper, 2010) and being located inside a volcano, Aso Farmland takes full advantage of these resources. As a somma caldera, the Central Cone is comprised of five peaks, including the active volcanic peak Nakadake. The magma chamber feeds into the underground hydrothermal reservoirs (approximately 50 m below surface) from which boreholes control the supply to the hot springs. Hot springs also naturally seep out on to the mountainside (Tsuji *et al.*, 2017). Engaging multiple senses, tourists experience a range of aromas from sulphur baths, to perfume and herb baths. The main *onsen,* Volcanic Hot Spring, is divided between men and women spas, each approximately 3300 square metres (0.8 acres) with parts of the *onsens* outside. The source of the hot spring from Aso Volcano contains magnesium, sodium and calcium, all known for being effective on the improvement of arteriosclerosis and skin problems, as well as having detox effects and stimulating relaxation. This mineral hot spring supplies the indoor baths and the outdoor baths but there are also steam rooms and dry saunas within the bath garden. There are more than 20 baths and saunas in the bathing facility. In the *onsen* facilities, no bathing suit or towels are allowed, as is the custom in Japan.

In addition to the variety of *onsen* baths, there are also 13 heated domed saunas (Earth power spas) lined with different herbs and minerals. Guests can lie in a heated dome covered with jade or amethyst, or with herbs such as lavender, Japanese mugwort, mint and cinnamon. These herbs, and minerals are selected based on the health effects of each ingredient in a heated environment. Alternatively, an 'ice' dome is another option where visitors experience frigid conditions.

Aso Farmland as a Disaster Relief Evacuation Centre

Having set out the tourist facilities, the chapter now turns to examine the stages that Aso Farmland went through as an evacuation centre which began soon after the 16 April Kumamoto Earthquake. The contribution made by Aso Farmland to the surrounding communities will be examined from both primary and secondary information. Primary information was obtained through an interview with the director of Project Planning and Communication at Aso Farmland, Ms Moriyama, interviews with taxi drivers as well as site visits where observations were made of damaged and newly constructed facilities. Secondary information was obtained from news programmes, documentary films, online blogs of disaster volunteers (Japan Emergency NGO and Saigai NGO Yui/Tom Maehara), Aso Farmland's Facebook and news releases and other social media publications, and publications by various government agencies and private organizations.

Initial phase – immediately after the main shock on 16 April 2016

The foreshock on 14 April did not cause any damage to Aso Farmland. It was 'business as usual' according to an Aso Farmland Facebook post-dated 14 April. However, it was the main shock on 16 April that devastated the Minamiaso area. While Mashiki town near the airport felt the earthquake as M_w 7, at Aso it was approximately M_w 6. The Minamiaso area was one of the worst damaged areas inside the caldera with over 100 landslides. Just before the main shock hit, Aso Farmland was about to open up its hot spring facilities for bathing to local residents whose houses were damaged by the foreshock.

At the time of earthquake, there were 200 guests at Aso Farmland and all were all unharmed in the earthquake proof domes. All tourism and hospitality operations had to be suspended and all of the tourists left the site the following day. The main facility buildings suffered only minor damage such as peeling paint and some cracked walls, however, there were broken underground water pipes that had to be repaired. Fortunately, damage to the underground gas pipes was very minor.

Disaster relief teams from around Japan responded very quickly. After the foreshock on 14 April, many municipalities convened emergency meetings on financial and humanitarian aid support for the earthquake-devastated areas (e.g. Cabinet Office, 2016b). This is a typical procedure in Japan where natural disaster relief efforts have become all too common. However, when the main shock hit the same area two days later, a variety of response teams and volunteers were immediately sent including: Defence Force troops, volunteers from Municipal and Prefectural Fire Departments, Police Departments, the Water Supply Department and Civil Engineering Departments. Many volunteers arrived in the affected areas in the early hours of the morning, stayed at evacuation centres or temporary bases so that they could begin excavation, debris removal and the delivery of relief aid as soon as dawn arrived on 17 April (e.g. see Japan Maritime Self-Defense Force website and various Japanese Municipality websites on Kumamoto Earthquake support).

Approximately 11,400 people in Minamiaso needed to evacuate not only because of the earthquake and aftershocks, but also because of the prolonged heavy rainfalls that followed the main shock. Many people stayed in their cars because of the frequent aftershocks. Sixteen designated evacuation centres/shelters (many in school sport gymnasia) accommodated 1036 people (Cabinet Office, 2016a). National and local governments asked Aso Farmland to take in evacuees to a maximum of 1,000 households knowing the domes at Aso Farmland were earthquake proof and safe. A few other hotels in the area also took in evacuees; yet in terms of the number of evacuees, Aso Farmland had the largest contingent.

Initially, Aso Farmland was to receive evacuees for approximately two months, until temporary housing could be built. The first group of evacuees arrived on 21 May and at the busiest time, up to 660 evacuees per day were on site. However, several months of heavy rainfalls after the main shock delayed the construction of temporary housing until October 2016. The last evacuee left the site in December. The majority of evacuees were not so fortunate to stay at hotel facilities as described by one taxi driver:

> Lucky evacuees could go home in 6 months or so. The majority of evacuees had to stay at school sports gyms for as long as a year. The temporary housing could not be set up that easily. In Minamiaso area there are

three or four, maybe five temporary housing complexes. Very few people stayed at hotels in the area but the majority were at school sports gyms. Neighbours provided soup kitchens and offered evacuees whatever they could. Sports gyms are not comfortable places. They just use partitions for very small living quarters. Babies and children cried especially at night. But there was no alternative – just persevere what we could do/have.

Originally Aso Farmland intended to rebuild its own tourism business first. The local municipal office however suggested that that the tourists would not come back to the region for some time, and this influenced the company's decision to prioritize the community and offer services to those in need.

Second phase – evacuees on site

The earthquake did not impact the main bath facility, however, the sauna and other bathing facilities were damaged including lifted floor tiles and cracks in the walls. The hotel wanted to open the main bath facility to locals as damage to water lines left many in the area without the option of taking a shower or bath. Since the main road in front of Aso Farmland was damaged, this was prioritized and was repaired with volunteer assistance in approximately two weeks. After fixing the underground water pipes (mid-April to mid-May), Aso Farmland began to take in evacuees to reduce the burden on other evacuation centres/shelters. This gave the locals a place to bathe and to live. The hotel also installed coin-operated laundry services in the accommodation zone for evacuees.

In an interview with the vice-director of project planning and communication at Aso Farmland, she explained how this project was financially supported. The local municipal office gave the hotel a per-person-per-day financial support, which included three meals a day. The money came through the Minamiaso municipal office that that was redirected from the National Emergency Budget. With the unexpected delays in temporary housing construction due to heavy rains, the evacuees had to be accommodated for longer than initially expected. This led to additional expenses for Aso Farmland and so when expenses exceeded the limited National Fund given to the municipalities, additional funding came from the local municipalities' budget. The Minamiaso municipality office set up an information desk at Aso Farmland to answer questions related to the earthquake, provide updates as well as all other necessary information for the evacuees.

Government agencies, communities and organizations from across Japan donated necessary supplies for the evacuees. Supplies were gathered at the Minamiaso municipal office and Aso Farmland sent a pick-up truck to the Municipal office for collection. A desk in the main lobby at Aso Farmland was set up so evacuees could get supplies (diapers etc.). The hotel covered extra expenses such as gasoline or employee's overtime.

During the school term, the lobby of the hotel became a gathering point for the children. In the morning the *randosel* (Japanese school backpacks) were all lined up as the children gathered in the lobby to catch the school bus. In the evening, children would do their homework at the desks in the hotel lobby.

Interviews revealed that the Aso Farmland employees found the evacuees on site to be very respectful and exhibited positive behaviour. The evacuees expressed gratitude to the hotel for opening their doors. Prior to coming to Aso Farmland, they were staying in designated evacuation centres/shelters where conditions were compared to 'refugee camps' – 'there were just partitions to mark individual spaces and no privacy'. They appreciated the space to walk around, no longer confined to a school gymnasium and they appreciated having their own homes (private spaces) as Aso Farmland's accommodation is in individual domes. Voices of the evacuees' included 'I can live again, feeling safe here', and 'We have privacy here' (Tora & Bell, 2016). One elderly woman who could not walk on her own before the earthquake relearned to walk again after coming to Aso Farmland. Walking on the premises from the accommodation zone to the restaurants and bath helped her strengthen her walking skills. The evacuees also dressed and behaved appropriately, so that when the tourists returned, they would feel that Aso Farmland is a tourist facility, and not an evacuation centre.

The company also had to deal with a variety of situations with their own employees. The overriding policy of the company was that no one would lose their job due to the earthquake. Many employees lost their homes and were living in evacuation centres and commuting to Aso Farmland while others lived in their cars. There were hundreds of aftershocks over several months, and some preferred their car so as to be ready to run at a moment's notice. The earthquake caused the closure of the local Tokai University campus in Tateno and many students were working part time at Aso Farmland. The university buildings and student residences were heavily damaged and the landslides destroyed the bridge outside the university entrance. With the university campus closed, there were no longer any university students working at Aso Farmland and this situation forced hotel staff to work multiple jobs.

While dealing with their own problems induced by the earthquake, the hotel employees faced a steep learning curve in dealing with the evacuees. The evacuees staying at the hotel were severely impacted both physically and psychologically. The terrified evacuees faced not only the mega-earthquakes and prolonged aftershocks, but also the loss of their homes, farmland, valuable possessions, if not family members. A large proportion of evacuees were elderly people and many lost their properties. They did not want to leave the town, but at the same time they were aware that they could not arrange for a new mortgage for a house due to their old age. Younger generations could move to other towns but they could not afford to rebuild their parents' houses. The earthquakes and landslides also destroyed farmland. The dikes between rice paddies were broken, landslides washed away small terraced rice paddies or debris, mud and silt from the landslides covered the fields, which made the area impossible to farm. The hotel staff had to deal with evacuees facing such desperate situations, although they were not trained as psychotherapists.

Third phase – rebuilding and recovery

The evacuees at Aso Farmland stayed from late May to December 2016. Simultaneously, Aso Farmland had to handle the reconstruction and repairs of their tourism facilities as well as be an evacuation centre. During the rainy season, western Japan was hit by unusually heavy rain for nearly 10 days (19–30 June); in addition, in the month of September, three typhoons (#12, 16 and 18) passed through northern Kyushu, including the Aso Area (Japan Meteorological Agency, 2016c, 2018c). These natural disasters delayed the recovery efforts and caused further damage by additional landslides and erosion, and flooding of rivers and towns. After shocks stronger than M_w 2 did not slow down for another year (see Table 7.2).

A message from Aso Farmland posted on the Sauna Spa Association online news release on 15 May 2016 announced that Aso Farmland would start receiving evacuees from 21 May; hot springs (indoor baths) are intact and open to disaster victims free of charge until 20 May; meanwhile outdoor baths and other sauna baths are under repair with broken floors, cracks on the walls and ruptured water pipes. Animals in the Petting Zoo are safe but have been evacuated to a safer place until the reopening. The company is targeting to reopen the facilities as early as the summer (August). An Aso Farmland online news release on 24 May 2016 explained in some detail the damage to the inside of restaurants, shops, fixtures, parts of lifeline infrastructure (e.g. water, electricity) and road surfaces on the premise.

Ms Moriyama tirelessly continued posting news, events, and progress reports on the Aso Farmland Facebook page from 16 April 2016 to date. This clearly reflects the importance of on-line communication strategies in destination image restoration theory as noted by Ketter (2016). Some Facebook members closely followed these postings adding their well wishes for the swift recovery of the hotel and the surrounding area. Ms Moriyama's on-line postings to many followers were real time (virtual) first-hand experiences of survival and recovery from the mega-earthquake.

Following the Facebook site of Aso Farmland, the 20 June 2016 entry describes entertainment for the evacuees including performances by the Japan Ground Self-Defence Force music band, the US Marine music band as well as volunteer Awa dancers from Tokushima Prefecture. At this point, most of the facilities on site were still closed for repair. Not only were evacuees on site, but also volunteer workers were also staying at Aso Farmland, offering physical labour, technical support, care for evacuees, and organizing events and activities for evacuees in collaboration with other evacuation sites and the Minamiaso Welfare division (see websites of Volunteer groups 'Aso Yomigaeri', 'Saigai NGO Yui', 'Tsumugiai', 'DRT Japan', etc.). The major recovery work such as rebuilding basic infrastructure, removing debris, clearing roads and fixing ruptured road surfaces, continued to October 2016.

Aso Farmland was acutely aware of the significance of reopening their business as soon as possible not only from a business perspective, but also for raising the morale of the devastated region. Minamiaso area is a known tourism destination and many tourism facilities in the area were damaged beyond repair. For instance, one taxi driver indicated there were about 20 B&Bs (guest houses) above on one of the cliffs but they were all put out of business after the main shock. It was a priority for the area to recover and re-establish tourism as soon as possible. Therefore, it was paramount for Aso Farmland to go back to 'business as usual', while continuing to serve the community. On 1 August 2016, Aso Farmland reopened to tourists while evacuees were still on site. In order to accommodate this, different dining areas were established for the two groups until eventually, all the evacuees were able to leave.

Final phase – over a year since the Kumamoto Earthquake

In June 2017, 14 months after the Kumamoto Earthquake, Aso Farmland completed repaving the roads within the property which had to be dug up to fix the water lines. However, in 2017 a two-day torrent of rainfall (5–6 July), two typhoons in July (#3 and 5), one typhoon in September (#18) and one in October (#22) passed through northern Kyushu. Typhoon #3 (30 June – 10 July) combined with extremely heavy rainfall brought new natural disasters in many places in northern Kyushu. The Aso Area was no exception and the excess water loosened land surfaces resulting in more landslides.

It is taking time for infrastructure recovery. The Government's assessment for full recovery of major road systems is three to five years. Japan Railway Kyushu started reconstruction of the Hōhi line in June 2017. On the Aso Farmland and Minamiaso official websites, it notes access to the area is via detours and nearest train station to Aso Farmland is inaccessible. Local residents are also frustrated by longer travel times on winding mountainous roads with one lane sections. When the authors visited the site in June/July 2017, many houses were still covered in blue tarpaulin sheets. A taxi driver explained that 'It took us (local people) one year to realize the entirety of what needed to be done. That's why reconstruction and recovery efforts are progressing so slow'. The next significant challenge is rebuilding the agricultural industry in the Aso area. Aso Farmland lost its own farms and hydroponics during the earthquake. Reaching out and reconnecting with the local farmers was of the utmost urgency. Many farmers lost their fields rice paddies and cattle to the earthquakes and landslides/debris avalanches. Kumamoto Prefecture is investing a lot of effort into the recovery of agriculture. Kumamon Bear (a Prefectural costumed mascot Black Bear) has been travelling to other parts in Japan promoting Kumamoto and its recovery projects, including agriculture and tourism.

As of June/July 2017 Aso Farmland was in the middle of considerable renovation and reconstruction. Many shops and restaurants and heated saunas described earlier in this chapter were closed. However, this is not due to the seriousness of the damage due to the earthquake but due the parent company's current policy. While the Japanese Government gave financial support to business owners for repairs, the owners of Aso Farmland decided that more than just repairs were needed. In their opinion, simply repairing the damage would not change the negative image of Aso after the Kumamoto Earthquake. Ms Moriyama stated that the owners:

> needed a more positive, more proactive and more attractive image of Aso Farmland … they wanted to renew the resort so the owners are putting a lot of money into resort redevelopment. They want to be known as a resort with new attractions and not just a repaired resort.

With echoes of 'Build Back Better', Aso Farmland is expecting to complete all the renovations and new construction in 2018.

Conclusion

The Aso caldera contains an active volcanic cone and people's concerns were about a possible eruption of Nakadake Mountain (volcano), but ironically, not about an earthquake. The 2016 Kumamoto Earthquake was different from any other mega-earthquakes in the past in terms of shallowness of epicentres, the long-period ground motion, extremely high number and long duration of aftershocks and probable intervention of seismic motions by the magma chamber. The disaster was both immediate and sustained and it is not surprising that there were no detailed disaster plans for this scale of disaster. Thousands were left homeless and school gymnasiums became the temporary home for many. Aso Farmland's dome shaped earthquake proof accommodations proved to be reliable structures, welcoming evacuees in this unfortunate incident. This case study illustrates that tourism and hospitality establishments can be a safe evacuation centres/shelters in time of need. In regions where disasters strike, visitor numbers will drop, leaving empty rooms that locals could potentially occupy if the situation is safe and stable. Aso Farmland also illustrated with some creative ideas, tourism establishments can simultaneously operate as a tourism facility and as a disaster relief centre. For example, through careful scheduling, group tours from South Korea and China returned to Aso Farmland as soon as the hotel reopened while it was an evacuation centre.

For this type of initiative to be successful there needs to be a tremendous amount of collaboration between local officials and the tourism industry. It calls for a great deal of sharing of information and resources especially funding from various levels of government. Aso Farmland's involvement in disaster relief services for the community was unquestionably a huge undertaking. Mr Masuda of Aso Farmland responded in an interview for an online news release (Tora & Bell, 2016):

> what is important is the connections; earthquake victims' helping each other is a connection. Tourist destinations and tourism establishments should be more connected [with each other and with tourists]. We have been working independently, like a dot. But learning from the invaluable lessons of supporting each other during the time of disaster, we need to change from being a dot to having a connection with each other ... to provide better tourist experiences.

Similarly, Ms Moriyama of Aso Farmland reflected on how interagency planning has changed after the 2016 Kumamoto earthquake.

> After the earthquake, Aso Farmland staff would go every day to the Minamiaso Task Force meeting to decide on what to do in terms of planning and policy. After the earthquake the community became as one. Before it was just communication between one agency to another, now there is more collaboration and the community worked together as one. They are rebuilding the community together.

It is clear that organizational learning has occurred not only at Aso Farmland but also with other agencies involved in disaster relief. New and stronger networks have been formed and Aso Farmland is 'Building Back Better'. Resilience has been strengthened and the community is better prepared. The accommodation centres need not only financial support and supplies but also for long staying evacuees, professional counselling services should be made available. This case also demonstrates the importance of online communication for destination image restoration. Based on the findings of this case study it is recommended that the tourism industry, and especially the accommodation sector be more involved in wider crisis management planning.

References

Aoi, S. (2016) *Jishin dēta kara mita Kumamoto Jishin [The Kumamoto Earthquake from Earthquake Data]*. Network Centre for Earthquake, Tsunami and Volcano (National Research Institute for Earth Science and Disaster Resilience), 24 April.

Aso Farmland (2014) *Kanren Gaisha (Kenkou no Mori Group) [Related companies (Healthful Forest Group)]* [online]. See http://www.asofarmland.co.jp/information/company/index.html#kenkoumori.

Aso Kumamoto Airport (2016) *2016/05/25 China Airline (Kumamoto-Kaoshun sen) no unkōsaikai nit tsuite* [25 May 2016 about resuming China Airline flights (Kumamoto-Kaoshun line) [online]. See http://www.kmj-ab.co.jp/news.html?st=65&caten=.

ASO National Park Healthful Forest (2013) *Aso Farm Land Guide Book*. Minami Aso, Kumamoto: Aso National Park Healthful Forest.

Becken, S. and Hughley, K. (2013) Linking tourism into emergency management structures to enhance disaster risk reduction. *Tourism Management* 36, 77–85.

Cabinet Office (2016a) *Heisei 28-nen Kumamoto-ken Kumamoto-chihō o shingen to suru Jishin ni kakawaru higai jōkyō tō ni tsuite – Hijō-saigai taisaku honbu,* [2016 Kumamoto Earthquake Summary – Disaster Management in Japan] [pdf]. See http://www.bousai.go.jp/updates/h280414jishin/pdf/h280414jishin_32.pdf.

Cabinet Office (2016b) *Heisei 28-nen Kumamoto Jishin, Jishin Gaiyou – Naikaku Bousai Tantō,* [2016 Kumamoto Earthquake Summary – Disaster Management in Japan] [pdf]. See http://www.bousai.go.jp/updates/h280414jishin/h28kumamoto/pdf/h280729sanko01.pdf.

Cooper, M. (2010) Volcano and geothermal tourism in Japan – examples from Honshu and Hokkaido. In P. Erfurt-Cooper and M. Cooper (eds) *Volcano & Geothermal Tourism Sustainable Geo-resources for Leisure and Recreation* (pp. 155–169). London: Earthscan.

Erfurt-Cooper, P. and Cooper, M. (2009) Introduction: Development of the health and wellness spa industry. In P. Erfurt-Cooper and M. Cooper (eds) *Health and Wellness Tourism* (pp. 1–24). Bristol: Channel View Publications.

Faulkner, B. (2001) Towards a framework for tourism disaster management. *Tourism Management* 22 (2), 135–147.

Ghaderi, Z., Som, A. and Wang, J. (2014) Organizational learning in tourism crisis management: An experience from Malaysia. *Journal of Travel & Tourism Marketing* 31 (5), 627–648.

Hall, C.M. and Mitchell, R. (2002) Tourism as a force for gastronomic globalization and localization. In A. Hjalager and G. Richards (eds) *Tourism and Gastronomy* (pp. 71–87). London: Routledge.

Hystad, P.W. and Keller, P.C. (2008) Towards a destination tourism disaster management framework: Long-term lessons form a forest fire disaster. *Tourism Management* 29 (1), 151–162.

International Civil Aviation Organization (ICAO) (2016) Assembly – 39th Session, Executive Committee: Quick response after earthquake in Kumamoto airport. *International Civil Aviation Organization Working Paper*. A-39/WP280 [pdf]. See https://www.icao.int/Meetings/a39/Documents/WP/wp_280_en.pdf.

International Research Institute of Disaster Science (IRIDeS) (2016) *2017 Kumamoto Earthquake Report* [pdf]. See http://irides.tohoku.ac.jp/media/files/earthquake/eq/2016_kumamoto_eq_report/2016_KumamotoEqReport_12-appx.pdf.
Japan Meteorological Agency (2016a) *Shingen Yōso (Shindo 5-jaku ijō)* [Elements of Epicentres (Magnitude 5- or stronger)] [pdf]. See http://www.data.jma.go.jp/svd/eqev/data/2016_04_14_kumamoto/i5.pdf.
Japan Meteorological Agency (2016b) *Chōshūki jishindō ni kansuru kansoku-jōhō (shikō)* [Observations on the long-period earthquake ground motion (trail)]. [online]. See http://www.data.jma.go.jp/svd/eew/data/ltpgm_explain/data/past/20160416012510/index.html.
Japan Meteorological Agency (2016c) *2016-nen (Heisei 28-nen) no taihū ni tsuite (Heisei 28-nen 12-gatsu 21-nichi)* [About typhoons in 2016 (21 December 2016)] [pdf]. See http://www.jma.go.jp/jma/press/1612/21d/typhoon2016.pdf.
Japan Meteorological Agency (2017a) *Omona Jishin no Hasseibasho no Shousai oyobi jishin Katsudō no Sui'i (2016-nen 4-gatsu 14-nichi – 2017-nen 4-gatsu 30-nichi)* [Details of main epicentres and changes of tremor activities (14 April 2016 – 30 April 2017)] [pdf]. See http://www.data.jma.go.jp/svd/eqev/data/2016_04_14_kumamoto/kouiki.pdf.
Japan Meteorological Agency (2017b) *Heisei 28-nen (2016-nen) Kumamoto Jishin (Heisei 28-nen 4-gatsu 14-ka 21-ji ~ Heisei 29-nen 7-gatu 11-nichi genzai Kishōchō Jishin-kazan bu)* [2016 Kumamoto Earth Quake (from 9:00 PM, 14 April 2016 ~) As of 11 July 2017 by Seismology and Volcanology Department, Japan Meteorological Agency] [pdf]. See http://www.data.jma.go.jp/svd/eqev/data/2016_04_14_kumamoto/yoshin.pdf.
Japan Meteorological Agency (2018a) *Chōshūki jishindō ni kansuru kansoku-jōhō (shikō) no happyōjōkyō* [Posting situation of observations on the long-period earthquake ground motion (trail)] [online]. See http://www.data.jma.go.jp/svd/eew/data/ltpgm_explain/data/past/past_list.html.
Japan Meteorological Agency (2018b) *Heisei 28-nen (2016-nen) Kumamoto Jishin no shindo 1 ijō no saidai shindo-betsu jishin kaisū-hyō* (Heissei 28-nen 4-gatsu 14-ka 21-ji ~ Heisei 30-nen1-gatsu 31-nichi 24-ji) [2016 Kumamoto Earthquake List of frequency of quakes stronger than M1 (9 PM, 14 April 2018 ~ 12 AM 31 January 2018)] [pdf]. See http://www.data.jma.go.jp/svd/eqev/data/2016_04_14_kumamoto/kumamoto_over1.pdf.
Japan Meteorological Agency (2018c) *Saigai o motariashita kishou jirei (Heisei gan'nen ~ hon'nen) [Examples of meteorological phenomena that caused disasters (1989~2018)]* [online]. See http://www.data.jma.go.jp/obd/stats/data/bosai/report/index_1989.html.
JR Kyushu (2018) *Unkōjōhō* [Train schedule information, 20 February] [online]. See http://www.jrkyushu.co.jp/UNKOU/mobileinfo.html.
Ketter, E. (2016) Destination image restoration on Facebook: The case study of Nepal's Gurkha Earthquake. *Journal of Hospitality and Tourism Management* 28, 66–72.
Lin, A., Satsukawa, T., Wang, M., Mohammadi Asl, Z., Fueta, R. and Nakajima, F. (2016) Coseismic rupturing stopped by Aso volcano during the 2016 Mw 7.1 Kumamoto earthquake, Japan. *Science* 354 (6314), 869–874. DOI: 10.1126/science.aah4629.
Mannakkara, S., Wilkinson, S., Willie, M. and Heather, R. (2018) Building back better in the Cook Islands: A focus on the tourism sector. *Procedia Engineering* 212, 824–831.
Medina-Muñoz, D. and Medina-Muñoz, R. (2013) Critical issues in health and wellness tourism: An exploratory study of visitors to wellness centres on Gran Canaria. *Current Issues in Tourism* 16 (5), 415–435.
Miyabuchi, Y. (2016) Landslide disaster triggered by the 2016 Kumamoto earthquake in and around Minamiaso Village, Western Part of Aso Caldera, Southwestern Japan. *Journal of Geography (Chigaku Zasshi)* 125 (3), 421–429.
Ministry of Land, Infrastructure, Transport and Tourism (MLIT) (2018) *Press Release: Private Sector Outsourcing of the Operation at Kumamoto Airport will start from*

April 2020 ~Toward accelerated recovery from the Kumamoto earthquake~ [pdf]. See http://www.mlit.go.jp/common/001217358.pdf.

Morakabati, Y., Page, S. and Fletcher, J. (2017) Emergency management and tourism stakeholder responses to crisis: A global survey. *Journal of Travel Research* 56 (3), 299–316.

NHK Kabun (2016) *Hijō ni tsuyoi Chōshūki jishindō, Kumamoto jishin de kansoku* [Very strong long-period earthquake ground motion observed in Kumamoto earthquake]. Science and Culture Blog [Blog]. See https://www9.nhk.or.jp/kabun-blog/200/245027.html.

NHK (Japan Broadcasting Corporation) (2017) 'Chokkagata jishin no aratanaru kyōi, Chōshūki parusu' [New threat of locally hit earthquake, long-period pulse]. *Ohayō Nippon*, 1 September [online]. See http://www.nhk.or.jp/ohayou/digest/2017/09/0901.html.

Orchiston, C. and Higham, J. (2016) Knowledge management and tourism recovery (de) marketing: The Christchurch earthquakes 2010-2011. *Current Issues in Tourism* 19 (1), 64–84.

Nishinippon Shinbun (2016) Minamai Aso no hinansho jōkyō (Kumamoto jishin) [Conditions of evacuation points in Minamiaso (Kumamaot Earthquake)]. *Nishi-Nippon Newspaper*, 17 April/18 May [online]. See https://www.nishinippon.co.jp/nnp/shelter_kumamoto/article/239032/.

Norimono News (2017) Minamiaso tetsudō zensen fukkyū ni sukunakutomo 5-nen, hashi kakekae ya ton'neru tanshuku mo. *Testudō* [Minamiaso railway full recovery requires at least 5 years, building new bridges, also shortening a tunnel.] *Railway*, 18 April [online]. See https://trafficnews.jp/post/68691.

Parsons, W. (1996) Crisis management. *Career Development International* 1 (5), 26–28.

Ritchie, B. (2004) Chaos, crisis and disasters: A strategic approach to crisis management in the tourism industry. *Tourism Management* 25 (6), 669–683.

Real Estate Japan (2016) *Dome Houses of Japan: Made of Earth-quake Resistant Styrofoam* [Blog]. See https://resources.realestate.co.jp/living/dome-houses-of-japan-made-of-earthquake-resistant-styrofoam/.

Scott, N., Laws, E. and Prideaux, B. (2007) Tourism crises and marketing recovery strategies. *Journal of Travel and Tourism Marketing* 23 (4), 1–13.

Schmidt, P. and Berrell M. (2007) Western and Eastern approaches to crisis management for global tourism: Some differences. In E. Laws, B. Prideaux and K. Chon (eds) *Crisis Management in Tourism* (pp. 66–80). Wallingford: CABI.

Tagami, I. (2017) Kumamoto kūko, kokusaisen V-ji kaifuku; Riyōkyaku kako saikō no ikioi [Kumamoto Airport shows a v-shape recovery; the highest number of passengers]. *Kumamoto Nichi Nichi Shinbun*, 18 December [online]. See https://this.kiji.is/315307762166711393.

Takayama, M. (2017) Dai 5-kai: Kumamoto jishin ni okeru menshin kōzō no kōka [Effects of seismic isolation structure during Kumamoto earthquake] *Takayama Mineo no hōmu pēji: menshin kōzō ni kansuru kenkyū, kenchiku kōzō-butsu no shindō mondai ni kansuru kenkyū* [Mineo Takayama's Home Page: Research on seismic isolation structure, Research on movement problems of architectural structures] [online]. See http://4menshin.net.

Telfer, D.J. (2015) Tourism and regional development issues. In R. Sharpley and D. Telfer (eds) *Tourism and Development: Concepts and Issues* (pp. 140–178) (2nd edn). Bristol: Channel View Publications.

The Mainichi (2016) 'Quake-proof, dome guesthouses offer comfort to Kumamoto evacuees'. *The Mainichi,* 10 July [online]. See http://mainichi.jp/english/articles/20160710/p2a/00m/0na/002000c.

Tokyo Metropolitan Government (2016) Dai 2-shō Heisei 28-nen Kumamoto Jishin no Gaiyō [Chapter 2, 2016 Kumamoto Earthquake Summary] *Kumamoto Jishin Shien no Kiroku ~ To no Bōsai Taisaku no Jikkōsei Kōjō ni Mukete~* [The record of Kumamoto

Earthquake Relief Support ~ towards the improvement of efficient disaster relief of Metropolitan Tokyo] 29 November 2016. Tokyo: Tokyo Metropolitan Government [pdf]. See http://www.bousai.go.jp/kaigirep/houkokusho/hukkousesaku/saigaitaiou/output_html_1/pdf/201601.pdf.

Tora & Bell (2016) *Daijishin o taenuita "Aso Farmland". Eigyō saikai made no kiseki <Kumamot-ken Minamiaso mura>* [Aso Farmland survived the mega earthquake. The path to the restart of business <Minamiaso village, Kumamoto prefecture>] [online]. See http://tora-bell.com/asofarmland/.

Tsuji, T., Ishibashi, J., Ishitsuka, K. and Kamata, R. (2017) Horizontal sliding of kilometre-scale hot spring area during the 2016 Kumamoto earthquake. *Scientific Reports* 7 (42947). doi: 10.1038/srep42947.

Yoshikawa, K. (2016) Minamiaso-mura de 1000-nin ga koritsu, Aso ōhashi ga houraku, dosha kuzure de dōro sundan (Kumamoto jishin) [1000 people stranded in Minamiaso Village, Aso ōhashi Bridge crumbled and landslides cut off the roads.] *The Huffington Post Japan*, 16 April [online]. See http://www.huffingtonpost.jp/2016/04/16/minami-aso-eathquake_n_9707176.html.

8 Handicraft Shopping Tourism after the Jogjakarta Earthquake: Recovery Network, Risk Perceptions and the Implications

Andri N.R. Mardiah and Jon C. Lovett

Introduction

The tourism sector is one of the engines of economic growth in many developing countries, and is highlighted as one of the strategies for reducing economic disparity (Jenkins, 1980). This is because the tourism industry is usually considered a labour-intensive industry which provides many opportunities for many local organizations and small businesses to grow (Dredge, 2006). Multiplier effects and tourism's links to many other sectors within an economy has been one of the sources of interest in the economic contribution of tourist shopping behaviour (Coles, 2004; Timothy, 2005), especially for micro and small retail enterprises that are important in developing economies (Tambunan, 2009; Hall *et al.*, 2016).

The disruption caused by a disaster results in fundamental problems for people's livelihoods and to many sectors within the affected regions (Phillips, 2009; Arendt & Alesch, 2015), including the tourism sector (Schwab, 2014). Although there have been many studies conducted on the economic impact of the disaster, research which focuses on micro and small businesses and their interconnections with the tourism sector remains relatively little explored. In particular, research focused on the how tourism businesses actually recovered, their expectations and the implications for future challenges.

On 27 May 2006 an earthquake measuring 5.9 on the Richter scale wreaked havoc on the Bantul region on the island of Java in Indonesia, causing a substantial economic downturn (International Organization for Migration, 2011). The earthquake resulted in the death of more than 57,000 people, more than 60,000 people injured, hundreds of thousands losing their homes and 2 million people, or half Yogyakarta province's population, being affected (Resosudarmo et al., 2008). At least 156,700 buildings were totally destroyed, and over 200,000 suffered varying degrees of damage (Resosudarmo et al., 2008). Damage from the earthquake to the economy and infrastructure in Jogjakarta and Bantul reached more than US$ 3 billion (BAPPENAS, 2006; Sekretariat JRF, 2011). This equals more than half of the Regional Gross Domestic Product (RGDP) of the preceding local government financial year (BAPPENAS, 2006). At the district level, 1328 craftspeople and 10,781 traders were identified as needing additional capital due to the loss of business assets. In addition, 29 traditional markets were damaged, causing a slowdown in local economic development (data recorded as per 7 June 2006, Bantul Regency, 2008: 116).

The Bantul and Jogjakarta provinces managed to recover within a period of two years (Tim Teknis Nasional, 2007), with good signs of recovery being identified much earlier (Bantul Regency, 2008; Government of Indonesia, 2010; Sekretariat JRF, 2011; International Organization for Migration, 2011). As a result, the experiences of Jogjakarta and Bantul may offer valuable lessons for other destinations affected by earthquakes. Thus, this research discussed the necessary recovery process after a disaster strikes and investigates tourism industry business communities, especially micro, small and medium enterprises (MSMEs), and their expectations for the future.

Methodology

The research focuses on Bantul Regency, especially on the sub-districts that have handicraft industries and shop clusters. The analytical unit used in this research is micro and small-scale business entities (MSMEs), almost all of them being home-based businesses. This reflects the fact that the local economy has been heavily influenced by small-scale economic activities. The sub-district samples were chosen by a purposive sampling method with sub-districts being selected which are located at high and medium risk level areas but which included more developed handicraft industrial clusters (sub-districts of Bantul Sewon, Kasihan, Bangun Tapan and Imogiri).

This study applies a mixed-methods approach of social networks (using UCINET and Netdraw), descriptive views (thematic content analysis) and questionnaires (descriptive statistical analysis) using data from the following: interviewing key actors; a questionnaire survey of 100 respondents in selected sub-districts; and a literature survey of the government policy documents and non-governmental disaster-recovery related project reports. The data collection was conducted in two phases: August–December 2015 and August–October 2016. The key actors' interviews were conducted during those two phases. The entire questionnaire was completed within the first phase. The questionnaire respondents are micro and small-scale business entities that were chosen by a random sampling within the selected sub-district samples.

Micro and Small Medium Enterprises in Shopping Tourism

There have been many studies exploring the relationships between shopping and tourism (Coles, 2004; Timothy, 2005; Swanson & Timothy, 2012; Azmi *et al.*, 2016). Buying souvenirs from handicraft shop clusters or shops near tourist destinations is, for many, an integral part of tourism. Swanson and Timothy (2012: 490) define a souvenir as a 'symbolic reminder of an event or experience', whereby it is a tangible proof of the intangible experiences of travel and/or leisure. Although integration is usually required with other tourist attractions (e.g. historical or natural attractions), the tourist shopping cluster can be a tourist destination in its own right (Timothy, 2005). Shopping tourism clusters, which are developed in a particular small geographical area, are sometimes referred to as tourist shopping villages (Getz, 1993). However, Getz (1993) further explains that such small cluster retailing activity could be within a small town or village, although usually located near tourist routes. Tourists may buy many other things besides handicraft; although unique handicraft items are one of the favourite souvenirs bought while travelling (Swanson & Timothy, 2012). The items chosen by tourists include arts and crafts, gemstones and jewellery, antique products, leather goods, housewares, pictures or statues of the landmark of tourist destinations, collectible items (e.g. mugs, key rings, fridge magnets, postcards), food originating from the destination area and local clothing (Timothy, 2005; Swanson & Timothy, 2012; Azmi *et al.*, 2016).

Most tourism is based on a network of small and medium-sized tourism enterprises and local organizations which provide all types of tourism products and services (Dredge, 2006). MSMEs contribute significantly to job creation, social stability, and the economic welfare of regions (Tambunan, 2009). The existence of MSME clusters within a region is also believed to encourage the formation of tourism patterns themselves, since clustering is believed to be a requirement for innovation and community capacity building (Dredge, 2006). The development of tourist destinations can also facilitate the growth of MSMEs as tourism can offers opportunities for new business start-ups including retail and shopping opportunities (Getz, 1993; Timothy, 2005).

Before and After the Earthquake: Handicraft Clusters and Tourism in Bantul Regency

Bantul Regency is part of the Special Province of Yogyakarta which covers four regencies and one city. Bantul Regency has an area of 506.85 km² and administratively it is divided into 17 sub-districts (*kecamatan*), 75 villages (*kelurahan*) and 933 hamlets (*dukuh*) (Bantul Regency, 2008). Based on population registration data, the population of Bantul regency has been increasing by 1.32% per year growing from 808,366 people in 2005 to 911,503 people in 2010 and to 917,511 people in 2015 (BPS Jogjakarta Province, 2016). The leading economic sectors in the Bantul regency from 2002–2006 included: the agriculture sector; manufacturing industry sector; trade sector, including hotels and restaurants; and the services sectors (Basuki, 2008). In 2005, 26% of the population worked in the agricultural sector, while 19% worked in industries and 21% in trade and commerce; by 2010 this combination had shifted toward trade, hotels and restaurants (26.54%), while the proportion of the population working at agricultural based activities decreased to 19.17% (Saputra & Rindrasih, 2012).

Geographically, tourism in Bantul Regency is closely linked to the City of Jogjakarta's tourism industry. This city is arguably the third most famous tourist destination in Indonesia after Bali and Lombok and is well known as a city of culture, heritage, education and natural beauty. Jogjakarta attracted on average 1,139,922 visitors per year during the period 2004–2008 (Dinas Kepariwisataan Jogjakarta, 2008) but this number does not include the many students from all over Indonesia due to the existence of many universities in this region.

When the earthquake struck Jogjakarta, MSMEs were the most severely affected, although at the same time they had the fewest resources able to be used to recover their businesses and livelihoods (Sekretariat JRF, 2011). As a result, the perceived impact on the local tourism economy was quite substantial, primarily due to the large number of MSMEs and handicraft clusters in Bantul Regency. There are 72 handicraft clusters in Bantul (Bantul Regency, 2008) which are not homogenous from one sub-district to another, in which most of the sub-districts already have their own featured or typical types of products. Some of the well-known tourist shopping platforms in Bantul Regency include the clusters of Kasihan (pottery craft), Manding (leather-based craft) and Imogiri (metal-based jewellery), as well as Wukirsari (batik craft). In Bantul Regency, the involvement of the village community in developing their clusters as tourism destinations is very important with the business incubation process being tiered upwards from the village community to the regency government level (Saputra & Rindrasih, 2012).

Based on observation, almost all the business entities within the handicraft industries in Bantul are small and home-based enterprises.

Most of them have a workforce of up to 20 people with the handicraft industry in Bantul Regency beings an important source of job creation in the region, employing around 60,000 workers (Bantul Regency, 2008). In addition, around 65% of the total handicraft exports in Jogjakarta Province are supplied by the industries located in Bantul Regency (Bantul Regency, 2008). After the earthquake in 2010, Bantul had 18,119 industrial entities, primarily consisting of small and medium enterprises (BPS Bantul Regency, 2011) with this number increasing to 20,423 industrial entities by 2015 (BPS Bantul Regency, 2017).

Analysis and Findings

This research is partly based on data collected from questionnaires. From 125 questionnaires collected, about 100 were usable with the number almost equally distributed within the sampling sub-district areas. Forty-nine percent of all the respondents are male, and 51% female. Most respondents (75%) were over 40 years old and the rest below that age (25%). The oldest respondent was 73 years old, and the youngest was 18 years old.

The majority of the business entities (65%) were established before the 2000s, with 51% being established during the 1980s or 1990s and approximately 14% before that era (i.e. from the 1950s to the 1970s). The business is dominated by micro and small businesses (76%), comprising 35% for micro-scale businesses and 41% for small businesses. Meanwhile, the rest are medium-sized businesses. Many of them have a workforce of up to 20 people (84%) with capital of no more than IDR 50 million (64%).

Those who are members of the clusters are generally craftsmen as well as sellers (i.e. doing production and marketing activities; 72%). Despite being small-scale businesses activities about 47% of them are export-oriented. However, their businesses are varied; most are dependent on their location in relation to the nearest crafts clusters. The composition of their core businesses is indicated in Table 8.1.

Table 8.1 Core business of respondents

Core business	%
Handicraft made from wood or bamboo	36%
Leather based crafts	20%
Batik crafts	16%
Pottery crafts	10%
Metal-based Crafts (e.g. silver, cooper, brass and the like)	9%
Others	9%

Recovery network in Bantul

Almost all respondents (96%) agreed that when the earthquake occurred in 2006, the disaster greatly affected their business continuity. From those who were affected, 56% of the businesses were closed down temporarily for one to three months, 22% for three to six months and the rest (22%) for more than six months. This situation was addressed through a series of policies, regulations and programme activities from many actors, such as basic living-needs assistance, housing reconstruction (either temporary or permanent), livelihood recovery and local economic development.

In highlighting the post-disaster programmes for MSMEs, the redevelopment of MSMEs in Indonesia was primarily managed by the Ministry of Cooperation and Small Medium Enterprises. In addition, the Ministry of Tourism and Bank Indonesia (i.e. the central bank of Indonesia) also participated in policy development. The local government of Bantul Regency approached MSME recovery with many programmes and activities focussed on local economic development (Saputra & Rindrasih, 2012). In addition to this, to promote shopping tourism, agencies such as the Cultural and Tourism Board and the Cooperation and SMEs Board at the local level were working hand-in-hand to provide assistance to the recovery of handicraft shopping tourism clusters in Bantul.

The Bantul government and the Cooperation and SMEs Board appeared as core meso-level actors in the recovery network. The main actor at the international level was an intergovernmental organization, the International Organization for Migration; on the national level it is Tim Teknis Nasional; and at the local level it was the Cooperation and SMEs Board. The most critical player in the network was the Cooperation and SMEs Board, which acted as a connector of many small sub-networks, including linking directly to the beneficiaries.

Using multi-dimensional scaling methods, the whole network can be divided into two big groups, implicitly named macro–meso level actors and meso–micro level actors. At the macro–meso level, the assistance of international agencies was recognized: the European Union, Asian Development Bank (ADB) and World Bank. BAPPENAS, local governments and donors (consisting of the European Union, the Netherlands, the UK, ADB, Canada, Finland and Denmark) created a multi-donor fund – the Java Reconstruction Fund (JRF), with the Government of Indonesia (represented by BAPPENAS), European Union and World Bank forming the joint steering committee. The project was started in 2006 and officially closed in 2012 after the project period was extended due to the Mount Merapi Eruption in 2010. The project was meant to back up government programmes (i.e. Tim Teknis Nasional); therefore, many of the activities were in line with the government's efforts. However, some of them were also a refinement based on the lessons taken from previous disaster governance (e.g. MDFF Aceh-Nias).

At the meso-micro level, the recovery activities were partnered with many stakeholders: among others, local government boards, Bank Indonesia, microfinance institutions and PT Permodalan Nasional Madani (PNM), as well as the private sectors within the platform of business investors and partners. Broadly speaking, the macro-meso level was dominated by the bureaucratic nuances of the public sector, including international agencies; while the meso-micro level was dominated by economic networking activities, including the private sector.

Recovery period

Interestingly, through the above observed network, the indication of the recovery was felt much earlier than expected (Table 8.2). Approximately 55% of the 96 respondents affected could restart a business normally and had begun to earn a profit after the first six months, 16% started between a six-month and one-year period and the rest (29%) starting after a year. This is in line with the achievements in housing rehabilitation and reconstruction during the first 18 months (Tim Teknis Nasional, 2007), which significantly fulfilled housing needs. Similar to this, reports from other agencies (Sekretariat JRF, 2011; International Organization for Migration, 2011) have also shown the same pattern.

Most of the production activities, management and marketing of MSMEs are conducted at home. MSMEs are also characterized by their reliance on local resources, private or family ownership, small operational scale, labour-intensive nature, non-formal skills and the ease of penetrating business sectors, as well as employing a minimum of or no promotion. Although most of the MSMEs are categorized as being under family-based ownership, some have professional management with legal permits and clear business structures. However, it is many of them are still managed in a fashion that often mixes up family or domestic needs into the formal business balance sheet. With these characteristics, MSMEs will face different difficulties between one another; however,

Table 8.2 The recovery period of business entities after the Jogjakarta Earthquake

Recovery period	% of MSMEs
More than two years	8.3%
More than one year	20.8%
From 6 to 12 months	15.6%
From 3 to 6 months	14.6%
From 1 to 3 months	40.6%

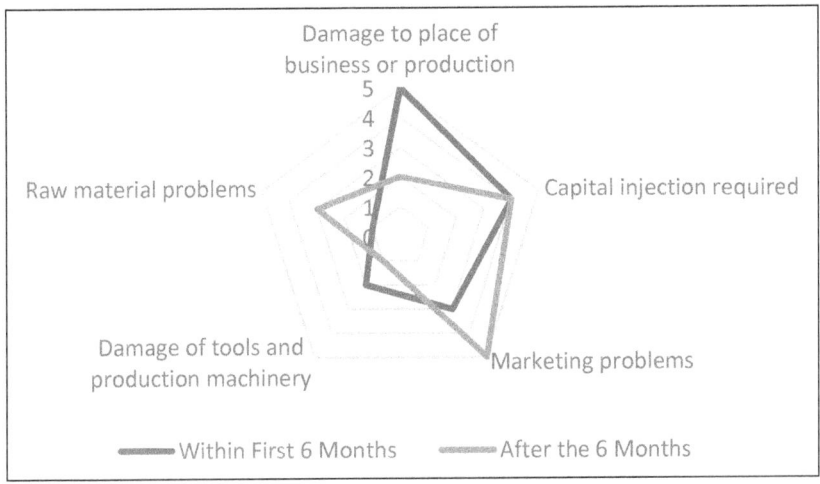

Figure 8.1 The most widely perceived business difficulties after the earthquake

the five most widely perceived problems during the recovery period are indicated in Figure 8.1.

A number of initiatives occurred to encourage business reconstruction and development. These are discussed under six main themes.

Affirmative and stimulant policy

Since many of the MSMEs in Bantul and Jogjakarta were home-based industries, damage to their houses definitely impacted on their continuity as small businesses. According to the respondents, the damage to their houses was the primary problem and presented the highest level of difficulty in business recovery (see Figure 8.1). The Government of Indonesia worked collaboratively with the local government, international agencies and NGOs, and made housing rehabilitation and reconstruction as the first priority after the emergency period was over. After six months, the housing problem had largely been resolved in most areas of Bantul as evidenced by a substantial decrease in housing demands. Six months after the earthquakes, respondents placed housing reconstruction as their fourth-highest priority (Figure 8.1).

However, the successful story of the affirmative policy of housing reconstruction was not followed by a smooth shift to policies that stimulated business recovery. In the second six-month period of recovery, the government failed to anticipate MSMEs' greatest difficulty of marketing. Marketing problems had actually been moderate in the first six months. However, while people still had their basic needs met by assistance from the government and or other parties, they did not demand too much of a solution. After the emergency period was over, aid became far less frequent, and crafts orders had been substantially decreasing since the earthquake (Sekretariat JRF, 2011). At that time, MSMEs realized that they had to speed up the business recovery in order to earn money at their normal levels.

The marketing problems were not only the result of infrastructure damage cutting distribution lines but also because they lost so many nearby customers due to the significant death toll. According to official government data, the total death toll in Bantul reached 5760 people (BAPPENAS, 2010). In addition to this, the MSMEs that relied heavily on the shopping tourism clusters (i.e. direct/offline marketing) suffered the hardest impact from post-disaster media coverage which resulted in the decreasing number of tourist visits to Jogjakarta.

The Bantul government claimed that small-scale manufacturing and handicraft products in Bantul Regency have been exported to the United States, Germany, Spain, the Netherlands, South Africa and Australia (Bantul Regency, 2008). However, this marketing development was not further encouraged due to the lack of promotion and the difficulties facing small and medium-sized craftspeople when conducting negotiations and transactions with potential buyers. To date, the export process has been conducted only by large companies or professional exporters; some of them even through the distribution channel of Bali tourism (Bantul Regency, 2008).

Aid disbursement and information

Most respondents said that they obtained assistance from the government. Besides government support, local and international NGOs, international agencies and private companies (through corporate social responsibility/CSR), as well as financial institutions also played an important role in supporting them in restarting their businesses. Interestingly, a number of respondents said that they also received assistance from the Indonesian Furniture Industry and Handicraft Association (ASMINDO), and personal donations and charity. These donations were conveyed through kinship relations and extended families, or even from unknown persons. Regardless of the source, these examples illustrate the important role that social networks and connections do play in the MSME recovery process.

In the beginning of the recovery phase, the communication of aid information regarding assistance from the Indonesian government and other agencies had been important for people. Later on, it became a crucial factor in eliminating any sense of unfairness between survivors as well as disbursing aid equitably. In the first three months, the situation was confusing due to poor management of information. Some people suspected there were unfair practices in the aid disbursement process. However, the government had decided on an order of priority for assistance for different groups.

The provision of aid also created unexpected problems. Some of the skilful senior craftspeople, supported by aid from government and non-government organizations, tended to build their own businesses, leaving the ones they used to work for temporarily disrupted due to a shortage of labour (Saputra & Rindrasih, 2012). After the earthquake, some entrepreneurs faced difficulties in hiring staff and skilful craftsmen since the new craftspeople had not received formal training, and most of the craftspeople learned their skills from senior workers or parents, without any formal training (Sekretariat JRF, 2011: 38). Indonesian government programmes and projects sought to provide training (e.g. design, product, business management and book keeping), for projects arranged by the International Organization for Migration (IOM) and the Deutsche Gesellschaft fur Internationale Zusammenarbeit GmbH (GIZ), that were financially supported by the Java Reconstruction Fund (JRF). Besides training, the livelihood recovery project of JRF covered access to microfinance, replacement of production tools, renovation of showrooms or workshops, access to markets, and capacity building (International Organization for Migration, 2011).

Financial access

Financial access was consistently the second highest priority of respondents. Three factors need to be noted in terms of empowering MSMEs via improving access to capital and finance: (1) the capital provision/aid must not cause any kind of dependency; (2) capital provision should occur through the creation of a system to gain access to financial institutions; and (3) the capital-allocation policy should not be misused. The objectives of facilitation of capital access through financial institutions are meant for MSMEs to gain experience in cooperating with existing financial institutions.

Besides giving financial access, MSMEs also demanded that the government pay more attention to all survivors' debt. Bank Indonesia (the central bank of Indonesia) followed this up by encouraging debt restructuring for Jogjakarta and Central Java earthquake survivors, with commercial banks restructuring 11,884 debts, and rural banks 4456 (BI Jogjakarta, 2007). In addition to this, the JRF agreed to increase financial access and to provide revolving funds for MSMEs. The project was partnered with local government, microfinance institutions and PT Permodalan Nasional Madani (PNM). Technical support was also provided to engage as many financial institutions as possible to ensure the coverage and sustainability of the programme. It was predicted this revolving fund would assist the MSMEs for at least until 2020.

Production facilities and infrastructure improvements

From the perspective of local economic redevelopment, infrastructure improvement was crucial. Accelerating productivity and business growth through capital access will have no significance if the product cannot be sold at the most appropriate price. Therefore, an important component in improving MSMEs' business after the earthquake was to improve infrastructure and transportation, thereby reducing production costs. In some cases, the craftsmen's villages were not heavily damaged, but the shop's cluster areas and the surrounding infrastructure were devastated, as in the case of silver craft production. In addition to this, interviews with MSMEs identified other potential threats to business continuity, with power outages being the most frequent and disruptive problem for operations and production processes.

Partnership and community engagement

Community engagement, based on the local wisdom of 'gotong royong, saiyeg saeka kapti' (mutual cooperation) (Tim Teknis Nasional, 2007), which was adopted as the value of partnership for sustainability (Sekretariat JRF, 2011), has been claimed as an essential element of successful recovery. For example, although the losses at the Kasongan pottery handicraft centre were estimated to be around 22 billion rupiah, the craftsmen worked hard to recover, and as a result, it was recorded that activities were relatively normal again by mid-2007 (Tim Teknis Nasional, 2007: 168–169). Furthermore, the spirit of togetherness was also found within the handicraft tourism cluster. The dynamic interactions between a shopping tourism destination and other kinds of tourism attractions strengthened both. As a result, local people began to believe that shopping tourism would provide them not only with sufficient income but also enable them to take on a profession, i.e. as an entrepreneur. Over time, the craftspeople or workers have been evolving and have become new entrepreneurs with tourism encouraging many new business players around the Bantul area.

In the context of MSMEs relating to large industries, business recovery was also driven by raw material assistance and production training: for example, the flour industry assisted micro-enterprises and home-based industries by conducting training on baking and food processing (Tim Teknis Nasional, 2007). Another example comes from within the framework of the JRF project that involved groups of women training in food processing in order to produce a variety of snacks and crackers from banana, cassava and peanuts. In addition to the food-processing skills, the women were also taught to create a brand, market it and find partners for marketing purposes (International Organization for Migration, 2011).

Local culture and networks

Most of the shopping tourism clusters in Bantul are based in one geographical area (i.e. they are village-based clusters), and their activities are dominated by cultural values of cooperation, social networks and trust. Although handicraft clusters have the same captive markets – the tourists who visit Bantul and Jogjakarta – they perceived that business competition remains relatively low. In a study of the connectedness and cooperation between the villages of Pundong and Kasongan, Saputra and Rindrasih (2012) found that most of the Pundong craftspeople produce a typical plain ceramic, but that those in Kasongan produce ceramics with various patterns. Many of the Kasongan craftspeople often buy the plain ceramics from Pundong and put a pattern or colours on them, e.g. woven rattan and banana stem bark patterns.

During the recovery period, the local government ran an aggressive campaign with the motto, 'Stand up on our own feet', in other words, the aid should assist communities so that they can help themselves (Tim Teknis Nasional, 2007). This message was meant to lighten the spirits of survivors who became depressed due to the aftermath of the disasters. Denis Nihill, the chief of Mission IOM Indonesia, stated, 'During these visits, I have also been pleased to note an increase of vibrancy and sense of optimism in the communities' (International Organization for Migration, 2011: 6). The government suggested people should not only be optimistic but also not over-dependent on any external aid. Some of the respondents said that they never received any assistance and relied solely on their existing savings and/or their own assets.

Strategies for the local economy
Re-developing the local tourism industry

The local economy was greatly affected by the earthquakes, especially in the regions that are heavily reliant on the tourism sector. Even if the tourism-related facilities can be quickly restored, the public perception of those destinations may not improve at the same pace. This is especially affected by media coverage and its influence on perceptions of a destination, so that a disaster-stricken area may not sufficiently attract visitors and investors (Schwab, 2014). In such a situation, Schwab (2014) argues that branding strategy is an essential component of economic disaster recovery in the most affected areas. However, the direct or indirect benefits of tourism do not always work as planned. In numerous cases, tourism development initiated from outside a region may overwhelm local people and their capacity to manage change (Nash, 1996).

One of the main challenges in Bantul Regency post-earthquake was the re-branding and promotion of the tourism sector. Bantul has long been known for its tourist attractions, cultural values, historical locations, and shopping tourism (Saputra & Rindrasih, 2012). Post-earthquake its capacity to generate employment is regarded as even more important given the boost the sector can provide to Bantul regional revenue (Bantul Regency, 2008). Local government is expected to show leadership in this area although government should anticipate the competing interests of local economic development with other concerns, such as social, cultural and environmental loss. Formulation of the future sustainable development of shopping tourism within the framework of regular village development planning processes (*musrenbang desa/kecamatan*) is essential and should be integrated with local business continuity planning to anticipate and manage future disasters.

Support to re start up business of the collapsed MSMEs

Even those businesses that do not receive direct damage may suffer extensive disruption from problems in infrastructure and/or the supply chain. Large companies within the community generally are already prepared with business continuity plans. Based on lessons learnt and perceived needs of MSMEs in Bantul-Jogjakarta, there are several strategies available for MSMEs to restart their business, including the settlement of non-performing loans for the affected enterprises and assistance in negotiating and restructuring credit obligations; improvement in the ability to access financing alternatives, including sharia schemes through micro finance institutions; capacity building in debtor business management, i.e. administration, bookkeeping and business management as well as other training and educational development opportunities, such as business continuity planning; promotional assistance; and the replacement of productive assets damaged by a disaster.

Conclusion

Most MSMEs are highly vulnerable due to the various 'shocks' that occur in the aftermath of a disaster. The Bantul and Jogjakarta earthquakes has a great impact on the financial capability of MSMEs. However, simultaneously, tourism MSMEs were also perceived by government as having an important role in revitalizing the economy after the disaster. Importantly for the process of post-disaster economic recovery, this case study reinforced the significance of affirmative policy from various levels of government, aid disbursement, financial access, partnership and community engagement, and local culture and networks as enabling factors. The case study also illustrates the way in which actors from different levels of governance worked together in the economic recovery process in the aftermath of disaster. Nevertheless, the MSME sector should be sustainably planned for the future and the development of MSMEs in Bantul will likely be driven by tourism both for its immediate economic impact but also because of its wider contribution to MSME development and recovery.

Acknowledgements

This chapter has benefited much from extensive discussion on literature, statistics and financial data from Noviyanto Utomo (Indonesia Financial Services Authority/OJK) and Purnama Alamsyah (Indonesian Institute of Science/LIPI). A special appreciation is dedicated to the Indonesia Endowment Fund for Education (LPDP) that funded this research.

References

Arendt, L.A. and Alesch, D.J. (2015) *Long-term Community Recovery From Natural Disasters*. Boca Raton, FL: CRC Press.

Azmi, A., Buliah, A. and Ismail, W. (2016) International tourist shopping styles in Kuala Lumpur. In *Proceedings of the 2nd International Social Science Conference 2016*.

Bantul Regency (2008) *Bantul Bangkit: Songsong Peradaban Baru*. Jogjakarta – Indonesia: The Goverment of Bantul Regency.

BAPPENAS (2006) *Preliminary Report of Damage and Loss Assessment*. Jakarta: BAPPENAS – Jogjakarta Province – Central Java Province.

BAPPENAS (2010) *Setelah Geulombang dan Lindu*. Jakarta – Indonesia: Sekretariat Tim Koordinasi Perencanaan dan Pengendalian Penanganan Bencana (P3B) – Direktorat Kawasan Khusus dan Daerah Tertinggal, BAPPENAS.

Basuki, A.T. (2008) Strategi Pengembangan Sektor Pertanian Pasca Gempabumi Kabupaten Bantul. *Jurnal Ekonomi dan Studi Pembangunan* 1 (9), 11–25.

BI Jogjakarta (2007) *2007 Economic Report on Yogyakarta Special Region*. Jogjakarta: Bank Indonesia.

BPS Bantul Regency (2011) *Bantul in Figures 2011*. Bantul – Jogjakarta: BPS.

BPS Bantul Regency (2017) *Bantul in Figures 2017*. Bantul – Jogjakarta: BPS.

BPS Jogjakarta Province (2016) *Daerah Istimewa Yogyakarta Province in Figures 2016*. Jogjakarta: BPS.

Coles, T. (2004) Tourism, shopping, and retailing: An axiomatic relationship? In A.A. Lew, C.M. Hall and A.M. Williams (eds) *A Companion to Tourism* (pp. 360–373). Oxford: Blackwell Publishing Ltd.

Dinas Kepariwisataan Jogjakarta (2008) *Buku Statistik Kepariwisataan Yogyakarta 2008*. Jogkakarta: Jogjakarta Province.

Dredge, D. (2006) Policy networks and the local organisation of tourism. *Tourism Management* 27 (2), 269–280.

Getz, D. (1993) Tourist shopping villages: Development and planning strategies. *Tourism Management* 14 (1), 15–26.

Government of Indonesia (2010) *Mitigasi: Menentukan Kembali Pengetahuan Kebencanaan Kita*. Jakarta – Indonesia: Kantor Staf Khusus Presiden – Republik Indonesia.

Hall, C.M., Malinen, S., Vosslamber, R. and Wordsworth, R. (eds) (2016) *Business and Post-disaster Management: Business, Organisational and Consumer Resilience and the Christchurch Earthquakes*. Abingdon: Routledge.

International Organization for Migration (2011) *Back on Our Feet: The Personal Stories of Micro and Small Enterprises Recovering from the Jogjakarta and Central Java Earthquake*. Jogjakarta – Indonesia: International Organization for Migration – JRF.

Jenkins, C. (1980) Tourism policies in developing countries: A critique. *International Journal of Tourism Management* 1 (1), 22–29.

Nash, D. (1996) *Anthropology of Tourism*. Kidlington, Oxford: Pergamon.

Phillips, B.D. (2009) *Disaster Recovery*. Boca Raton, FL: CRC Press.

Resosudarmo, B.P., Sugiyanto, C. and Kuncoro, A. (2008) *Livelihood Recovery after Natural Disasters and the Role of Aid: The Case of the 2006 Yogyakarta Earthquake*. Working Papers in Trade and Development No. 2008/21. Canberra: The Arndt-Corden Division of Economics Research, School of Pacific and Asian Studies, ANU College of Asia and the Pacific, Australian National University.

Saputra, E. and Rindrasih, E. (2012) Participatory planning and village tourism SMEs: A case study of Bantul Regency, Yogyakarta, Indonesia. *Geografia. Malaysian Journal of Society and Space* 8 (7), 54–64.

Schwab, J. (2014) *Planning for Post-Disaster Recovery: Next Generation*. Chicago: APA Planners.

Sekretariat JRF (2011) *Terus membangun dari Kesuksesan: Secara Efektif Menanggapi Beragam Bencana*. Jakarta – Indonesia: Sekreatriat JRF.
Swanson, K.K. and Timothy, D.J. (2012) Souvenirs: Icons of meaning, commercialization and commoditization. *Tourism Management* 33 (3), 489–499.
Tambunan, T. (2009) *SMEs in Asian Developing Countries*. Basingstoke: Palgrave Macmillan.
Tim Teknis Nasional (2007) *Membangun Daerah Rawan Bencana dengan Kearifan Lokal*. Jogjakarta – Indonesia: Tim Teknis Nasional – Provinsi Jogjakarta – Provinsi Jawa Tengah.
Timothy, D.J. (2005) *Shopping Tourism, Retailing, and Leisure*. Clevedon: Channel View Publications.

9 Earthquakes, Psychological Resilience and Organizational Resilience: Tourism Entrepreneurs in Kaikōura, New Zealand

Shupin (Echo) Fang, Girish Prayag and Lucie K. Ozanne

Introduction

Kaikōura is a small town in the South Island of New Zealand. A quarter of the jobs in this town come from the tourism sector and 34.1% of its GDP is related to international tourism spend (Statistics New Zealand, 2017). At 12.02am, on Monday 14 November 2016 NZDT, a magnitude 7.8 (Mw) earthquake struck the town. The earthquake caused two deaths and generated over NZ$900 million in insurance claims. In addition, State Highway 1, which is the main coastal road that connects several towns in the South Island such as Picton, Kaikōura and Christchurch, and which carries large amount of freight traffic between the north and south islands of the country, was severely damaged and partly closed for 13 months, along with the rail line (Bayer, 2017). Due to the road and rail closure and the negative media attention, Kaikōura experienced a sharp decrease in both tourism numbers and expenditure. Both expenditure from domestic and international tourists dropped by NZ$21 million from November to December 2016, with international spending dropping to zero in the first five weeks following the earthquake. Domestic tourism spend dropped by 7% on forecasted expenditure (Ministry of Transport, 2017). The town is heavily dependent on the tourism industry for its survival (Ministry of Business, Innovation & Employment (MBIE), 2017) and it was estimated that the percentage loss to business operability in the first week following the earthquake was 75% but this decreased to a 35% loss after six months (Ministry of Transport, 2017). Therefore, Kaikōura offers an interesting case study to understand how entrepreneurs in the tourism industry utilize psychological resilience to achieve positive business outcomes.

The impacts of disasters on tourism demand (Cró & Martins, 2017) and expenditure (Prayag et al., 2019) as well as on tourism businesses (Orchiston, 2013) have received increased research. Building resilience can be an effective way for tourism businesses to prepare for and respond to disasters (Biggs et al., 2012; Dahles & Susilowati, 2015; Jiang et al., 2019; Orchiston, 2013; Orchiston et al., 2016). For example, Jiang et al. (2019) stress the importance of building dynamic capabilities through a process of routine transformation, resource allocation and utilization for tourism businesses to respond to disruptive changes. At its core, resilience can be described as the capacity of a social-ecological system to withstand both sudden and incremental changes, allowing the system to adapt to the changed reality (Hall et al., 2018). The terms organizational, business and enterprise resilience have been used interchangeably to describe the resilience of businesses (Hall et al., 2016a; Hall et al., 2018). With the exception of the study by Prayag et al. (2020), which examined the influence of both psychological resilience and employee resilience on organizational resilience, the tourism literature remains thin with respect to the psychological resilience of tourism entrepreneurs and how it influences the way they cope and respond to disasters.

Psychological resilience has been defined as the positive psychological capacity for an individual to rebound or bounce back from adverse situations (Luthans, 2002). De Vries and Shields (2006) suggest that psychological resilience is developed through life experiences rather than being an inborn personality trait, and that psychological resilience identified among the New Zealand small and medium owner-operators positively contributes to business outcomes. Surprisingly, the literature at the intersection of entrepreneurship and resilience remains thin and fragmented (Korber & McNaughton, 2018). In times of adversity, entrepreneurs can benefit from being resilient as it allows them to not only adapt to unexpected changes more rapidly but also to capitalize on new entrepreneurial opportunities (Bullough & Renko, 2013; Duchek, 2018; Korber & McNaughton, 2018).

The purpose of this chapter is to illustrate how the psychological resilience of tourism entrepreneurs has a positive influence on organizational resilience. The chapter is based on qualitative research that was conducted in two phases. In phase one, the lead author attended the 'Kaikoura Challenge' in May 2016, which was sponsored by the University of Canterbury and the New Zealand Transport Agency. The 'Kaikoura Challenge' allowed university students to participate in a workshop to generate potential recovery strategies for small businesses in Kaikōura. The researcher as participant observer listened to the stories of small business owners in a local panel which comprised mainly tourism businesses. The one-day workshop also involved a brainstorming session that was aimed at identifying business opportunities and strategies that would enable the small town to recover from the earthquake. The second phase of the study involved semi-structured interviews with owner-operators of 18 tourism businesses, 8 males and 10 females, ranging in age from 25 to over 55 years old. Following Tongco's (2007) suggestion, a purposive sampling strategy was adopted, which allowed the researcher to deliberately choose participants who were knowledgeable about the topics. From the initial participants identified from the 'Kaikoura Challenge', additional participants were identified through snowball sampling. Field trips for data collection started in September 2017, ending in October 2017. The chapter focuses on reporting the findings from phase two of the study. The data saturation technique was employed for the sample size and the data recorded, transcribed and analysed using thematic analysis (Braun & Clarke, 2006).

Next, the chapter reviews the literature on the impacts of earthquakes on tourism businesses and small businesses in particular. This is followed by a review of the literature on psychological resilience and organizational resilience. Thereafter, the themes identified from the data analysis are presented. The chapter concludes with a discussion of how psychological resilience positively influence organizational resilience of tourism businesses.

Earthquakes, Tourism and Small and Medium Enterprises (SMEs)

Earthquakes are extremely unpredictable (Orchiston, 2012; Tsai & Chen, 2010), and can cause large casualties, significant damage to infrastructure and post-traumatic stress disorder (PTSD) for communities (Fan et al., 2011; Hall et al., 2016b; Satake, 2014). Earthquakes can also have severe impacts on tourism destinations and the tourism industry (Mazzocchi & Montini, 2001; Wang, 2009). New Zealand tourism businesses, in particular, are exposed to high seismic risk due to their geographic location (Orchiston, 2013). The New Zealand tourism industry, like many other industries, is dominated by SMEs (Ministry of Economic Development, 2011). Studies have shown that SMEs, when compared to large businesses, are mostly ill prepared for the aftermath and the recovery process following a disaster (Cioccio & Michael, 2007). However, SMEs often have their own path to get through natural disasters and positively recover from the loss (Biggs et al., 2012; de Vries & Hamilton, 2016). De Vries and Hamilton (2016) suggest that the 2010–2011 Canterbury earthquake sequence provided some small business owners with numerous opportunities to change their businesses for the better through a process of changing their operational business model, ownership structures, staffing strategies, mission statements and growth strategies.

Several researchers have call for more studies examining how tourism SMEs cope and respond to disasters (Biggs et al., 2012; Orchiston, 2012; Hall et al., 2018). A study on the 1994 Northridge earthquake in the Los Angeles, showed that among the 39,000 businesses impacted by the natural disaster, smaller businesses were in the worst-off group (Tierney, 1997). Cioccio and Michael (2007) studying the case of the 2003 bushfires in northeast Victoria in Australia, concluded that this natural disaster caused over 1000 small tourism businesses to have no revenue base. However, resilience built through accumulated experience enabled them to recover from the disaster. Biggs et al. (2012), focusing on the recovery of reef tourism businesses in Thailand, found that small businesses were more confident about their financial conditions after natural disasters than formal enterprises and reported higher levels of social capital. The three most important survival factors reported by these informal enterprises were commitment, support from government and NGOs and the ability to rely on a second source of income. The relationship between financial capital, social capital and resilience is implied in existing studies on SMEs. In the entrepreneurship literature, it has been argued that resilience is a poorly defined concept as it has a connotation with a wide range of other concepts such as success, survival, persistence and optimism (Korber & McNaughton, 2018). Entrepreneurs tend to be a heterogeneous group when it comes to their optimism, confidence and the identification of entrepreneurial opportunities (Shepherd et al., 2015). The term entrepreneurial resilience has, thus, been used to describe both individual resilience and the resilience of the entrepreneurial firm, requiring further clarification on how one affects the other (Korber & McNaughton, 2018).

From Psychological Resilience to Entrepreneurial Resilience and Organizational Resilience

The psychological resilience of entrepreneurs has received minimal attention in the psychology, entrepreneurship and tourism literatures. While some describe this as entrepreneurial resilience (Duchek, 2018), others call for a more dynamic conceptualization of doing resilience and the role entrepreneurship plays in shaping the positive trajectory of socioecological systems (Korber & McNaughton, 2018). Psychological resilience in the workplace is a well-studied topic (Avey et al., 2008; McDonald et al., 2016; Youssef & Luthans, 2007) along with how individuals cope with disasters (Mann et al., 2018; Smith et al., 2017). However, these two distinct bodies of knowledge and the entrepreneurship literature have not been well integrated to understand how entrepreneurs use psychological resilience as a resource to bounce back from adversity and how they apply this resource to achieve positive business outcomes (Duchek, 2018; Korber & McNaughton, 2018). The terms individual and psychological resilience have been used interchangeably (Hall *et al.*, 2018) to refer to the ability of individuals to 'rebound or bounce back from adversity, conflict, failure, or even positive events, progress, and increased responsibility' (Luthans, 2002: 702). This definition expands the traditional notion of the resilience of individuals from a psychosocial perspective to a developmental perspective (Masten, 2001), which so far is lacking in the entrepreneurship literature (Korber & McNaughton, 2018).

Historically, research on psychological resilience has been carried out in three waves. The first wave, focuses on the specific characteristics that make people resilient (e.g. Aroian & Norris, 2000; Christiansen *et al.*, 1997). This approach has dominated the few existing studies in the entrepreneurship literature that attempt to understand whether entrepreneurial resilience stems from personal characteristics of the entrepreneur (Korber & McNaugton, 2018). Resilient entrepreneurs are thus portrayed as individuals who thrive despite restrictive social, cultural and political norms or adverse conditions such as disasters (Loh & Dahesihsari, 2013). The second wave studies the process of obtaining such resilient characteristics (e.g. Compas *et al.*, 1995). Unfortunately, the entrepreneurship literature remains silent on how both individual and organizational actors utilize this innate resilience to cope with disasters (Korber & McNaughton, 2018). The third wave focuses on the motivational aspect of resilience, which is beneficial for generating personal strength for those in the organization (Richardson, 2002). This has implications for entrepreneurial behaviour fostering organizational resilience and also for entrepreneurial firms to foster the resilience of communities, regions and economies (Korber & McNaughton, 2018).

It largely remains unclear in the entrepreneurship literature what the term entrepreneurial resilience actually means and which factors help to enhance this capacity (Duchek, 2018). Depending on whether resilience is viewed as trait based, that is, an inborn human trait, then resilient individuals will display confidence, positive emotions and optimism. They will also exhibit characteristics of purposefulness through self-control and conscientiousness. Such individuals would have the ability to improvise and thus be adaptable but also sociable (Cooper *et al.*, 2013). De Vries and Shields (2006) describe entrepreneurial resilience as a collection of behavioural characteristics and can be seen as a process outcome rather than a personality trait. This outcome based perspective suggests that the resilience of entrepreneurs can be developed, managed and enhanced if they engage in business development and resilience training, build social capital through developing networks, are active in entrepreneurial pursuits, and actively seek objective, critical and encouraging feedback on their entrepreneurial decisions (Bullough & Renko, 2013). Following the Canterbury earthquakes of 2010 and 2011, entrepreneurs who showed positive attitudes towards the recovery were the most likely to change their existing business model and they tended to cope better post-disaster (de Vries & Hamilton, 2016). Bullough and Renko (2013) demonstrate that the combination of self-efficacy and psychological resilience provides individuals with more entrepreneurial power, especially under severe conditions, allowing them to identify new business opportunities.

There is a general agreement on the developable nature of psychological resilience that its development in the workplace should focus on reducing risk, building resilience strengths and utilizing effective adaption processes (Masten et al., 2009). However, others have suggested that psychological resilience is a component of psychological capital, which also includes optimism, hope and self-efficacy (Luthans et al., 2007). What distinguishes psychological resilience from other components of psychological capital is that: (i) it takes both proactive and reactive measures in the face of adversity, thus allowing the individual to both rebound and grow beyond the equilibrium point; (ii) the proactive nature of psychological resilience is the crucial characteristic that allows resilience to be distinctive, which stands for the ability to overcome, bounce back and further pursue new knowledge and meaning in life; and (iii) it combines assets and risk factors in an interactive pattern, and goes beyond the additive sum of an individual's assets and risk factors (Luthans et al., 2007). Given that entrepreneurial business decisions and outcomes are strongly influenced by the characteristics of the entrepreneur, it can be argued that psychological resilience of the entrepreneur should positively impact decision making in small tourism enterprises.

Studies in the workplace have shown that psychological resilience has positive effects on job performance and work happiness (Youssef & Luthans, 2007). Wang et al. (2018) showed that all components of psychological capital, including psychological resilience, had a positive impact on the job satisfaction of entrepreneurs. Entrepreneurs through their higher work engagement tend to score higher than employees on job satisfaction but they can also suffer the most from anxiety and depression when work-related issues impact business and personal outcomes. In a study of entrepreneurs, Duchek (2018) found that extreme work behaviours associated with over-competitiveness, passion and extraordinary persistence, blind faith in what they are doing and an ability to work harder than other people were the source of their entrepreneurial resilience. Resilience is necessary to overcome entrepreneurial challenges and to achieve financial success (Korber & McNaughton, 2018). Ayala and Manzano (2014) show that hardiness, resourcefulness and optimism have predicting power on the success of entrepreneurs and resilience can be seen as a real personal growth strategy for entrepreneurs.

The same positive outcomes of building psychological resilience in the workplace should, therefore, also be applicable to SMEs, where the resilience of both the entrepreneur and employees should positively impact post-disaster business recovery. However, there seems to be an implicit assumption that entrepreneurial resilience contributes to higher levels of organizational resilience (Korber & McNaughton, 2018). Developing psychological resilience can reduce personal vulnerability (Jackson *et al.*, 2007) and improve the work environment for all employees (McDonald *et al.*, 2016) but these do not automatically translate into the resilience of the business. Psychological resilience must translate into innovation and creativity, which then impacts positively organizational resilience. Dahles and Susilowati (2015) found that in micro-tourism enterprises, innovation, creativity and the ability to find alternative resources were critical for their resilience. Orchiston *et al.* (2016) found in businesses impacted by the Canterbury earthquakes, some level of pre-disaster planning, as well as an organizational culture that encouraged everyone to think outside of the box. In addition, collaboration with others and innovation were essential to build organizational resilience. The entrepreneurial resilience literature argues that the ability to act entrepreneurially, which emerges from an entrepreneurial mindset or entrepreneurial behaviour from founders and employees facilitate organizational resilience. However, what these constitutes are left unclear (Korber & McNaughton, 2018). Prayag *et al.* (2020) show that psychological resilience only impacts organizational resilience through employee resilience in small tourism firms.

Findings

The findings from analysing the transcripts thematically are presented under four themes. The first theme identifies psychological resilience as more a developmental process than a trait in entrepreneurs. The second theme suggest entrepreneurial experience as a source of psychological resilience. Psychological resilience developed through previous experience could positively contribute to resilience of the business. The third theme suggests that social capital facilitates psychological resilience. Resilient owner-operators were more likely to rely on social capital in the post-quake recovery period. The last theme discusses psychological resilience as a catalyst for business change. Resilient owner-operators saw the Kaikōura earthquake as a catalyst for making changes in their business that could eventually enhance the adaptive capacity of the business.

'Psychological resilience: Process versus trait'

Most participants viewed their own individual resilience as something they have developed, which conforms to the process view, suggesting that it is a developable capacity for people to rebound from adversity (Luthans *et al.*, 2007). It was mentioned by some participants that the things they had gone through, especially the 2010–2011 Christchurch earthquake sequence, helped them to be resilient as suggested in the quote below.

> I've also gone through the Christchurch earthquake … So, I think knowing what has happened in the past and roughly what to expect and what to do, because we've been through it before, it did help … But most of the time, if you have experienced certain things, you can be, for instance, aware of the consequences, or whatever of actions. And, more importantly, inactions. (Participant 9)

The data also suggested that psychological resilience was closely related to an individual's upbringing. This is similar to Duchek's (2018) study that found situational factors related to the entrepreneur's upbringing such as parenting style, assigning children responsibilities and challenging tasks, were formative to their resilience. Previous research recognized the contributing role of family in building psychological resilience (Mzid et al., 2019). The first quote below shows the developmental process of psychological resilience through learning from parents. The second quote shows the perceived developmental process of psychological resilience through being a part of large families.

> Well, it's probably, [developed] … well, it's probably … [upbringing]. I'd say that my parents were both quite resilient and that's probably why I am … Yeah, my dad's very … practical and knows how to fix a problem, and my mum's not very emotional (laugh) … So, I probably got it from both of them. (Participant 8)

> Well, both myself and my husband come from big families, so, we have to be resilient in our families (laugh) … you have to find your way through. So, I suppose that was building up as from childhood, really. As to … where we stood in the family, basically. (Participant 14)

However, only two participants suggested that psychological resilience was an inborn trait. One participant believed that he was not a 'panicker' and that resilience emanated from his 'DNA', as suggested in the quote below. The other participant expressed that resilience was something 'inborn and a family thing'. Interestingly, these two participants operated larger businesses compared to others. They seem to attribute the ability of their business to bounce back from the earthquake was dependent on their leadership role that stemmed from their inborn resilience trait. To some extent, this gives credence to the research stream on entrepreneurial resilience that focuses on trait aspects (Loh & Dahesihsari, 2013).

> So, from a resilience perspective, it is part of our DNA that we are not panickers … I do not believe that it is something that you can teach someone. It is about how you were built. You are either … not saying you are a panicker, but you are someone that will see a situation, and

will assess it very quickly ... make a decision and move. So, that's about, that is about resilience. And, some people, you know, not criticism on people, but, [they] can't deal with that, it's hard for them to deal with. (Participant 10)

'Psychological resilience: Entrepreneurial and disaster experience'

People who have previously experienced natural disasters are generally more resilient (Ferraro, 2003). Psychological resilience can be enhanced by previous experience in similar events. Participants who perceived themselves as resilient knew what to expect from the earthquake, were more prepared and had reasonable plans in place, as illustrated in the following quote:

In terms of things happening like earthquakes and fires etc., we have plans in place for that [planning strategies]. With the earthquake, my owner actually lost one of her hostels in the Christchurch earthquake ... She spent four years going through the process of cleaning that up, then getting the insurance and moving on for her life [previous experience]. And then, one day, she just bought another one, and kept going. So, we are very aware [situation awareness] of what's gonna happen [because of the earthquake].' (Participant 7)

As suggested by Ayala and Manzano (2014), resilience is the result of an interaction between entrepreneurs and their surrounding environment. This argument resonates with the present research and suggests that not only does previous interaction with the environment (i.e. previous experience in dealing with earthquakes) enhance psychological resilience but that participants also capitalized on their resilience to improve business performance. They tended to have pre-disaster plans in place, have more situation awareness and were more aware of the adaptive capacity of their business. Planning strategies and situation awareness capacity are key components that contribute to organizational resilience (McManus *et al.*, 2008).

'Psychological resilience: Social capital as an enabler'

As suggested by Aldrich (2012) and Ozanne and Ozanne (2016), social capital is the networks and resources built through connections to others and it is beneficial for post-disaster recovery. The data suggests that participants who displayed stronger psychological resilience relied on social support from family, friends and co-workers, and were also keen on doing things for others. More importantly, they seemed to be effective at establishing business partnerships and valued the importance of building networks. In this way, the results seem to suggest that social capital is an important enabler of the psychological resilience of entrepreneurs. The following quote provides a good example to illustrate this finding:

I do have some very good mentors, and people who I talk to ... well, not much business mentors, that he's a very successful business man, and he has been able to teach me through things [social referencing, process view of resilience] ... so, I tried to surround myself with the people who were positive [social support], and could see, you know, that life is a very, very long tunnel. But, if you could see that something happens and comes closer, and that's what's been happening ... you go out and you help other people ... By [doing something for somebody else], that helps you to recover [resilience] ... Because I shop locally most of the time, so, I try to get most of my suppliers locally. I've got a really good relationship with the local business people [networks] ... It's (the business) been resilient [organizational resilience], because we had a lot of help [network]. (Participant 4)

Effective business partnerships and networks are critical capabilities informing organizational resilience (Orchiston et al., 2016). Prayag (2018) argued that social capital has an important role to play in enabling organizational resilience. Hence, it can be argued that social capital acts an enabler of psychological resilience, which in turn positively impacts organizational resilience. In family firms, Mzid et al. (2019) found that social capital enhances the small firm ability to absorb shocks, reallocate existing resources and internalize practices that allowed them to cope with disturbances. The trust based networks of entrepreneurs allow them to cope with the changed reality (Danes, 2013).

'Psychological resilience: Catalyst for business change'

Resilient tourism entrepreneurs in this study viewed the earthquake from the unique angle that it is a catalyst for upgrading and changing their business. Not only did they bounce back from the negative impacts of the earthquake on themselves, but they also persevered to support their business. More importantly, they proactively found new paths and opportunities for the business post-disaster. This aligns with previous studies suggesting that small business owners can perceive a disaster as the catalyst for changing the existing business model (de Vries & Hamilton, 2016). This was perfectly demonstrated by those tourism SME owners who displayed the so called entrepreneurial mindset. For example, given that domestic tourists were experiencing significant difficulties to get to Kaikōura, one entrepreneur who relied heavily on this market changed the target market for his business to Chinese tourists who were less impacted by the uncertainty of the road closures to sustain revenue.

Another business owner who experienced difficulties due to geological damage in the area that isolated the business did not give up operating the business but instead started earthquake tours, telling stories about his experience of the earthquake. More importantly, previous studies have shown that successful entrepreneurs can utilize their resilience and are not easily knocked down by turmoil and adverse events. Entrepreneurial resilience enables entrepreneurs to persevere in the face of adversities and find new paths for the business rather than closing the business down (Bullough & Renko, 2013). Both the literature and the present research show that personal resilience enables the participants to persevere, continue, upgrade and even change the existing business model in the face of adverse events as shown in the quotes below.

> No, I've never [knocked down by events]. I didn't feel knocked down by the earthquake [psychological resilience]. It's the earthquake, with its repairs, it's also gonna stabilize our growth. (Participant 12)

> We now realize that we have to source our customers from a different country, like, Chinese nationals. I don't think you can call the Chinese market an emergent market anymore, you know. The New Zealand Tourism Board, for example, and that's been going on that for about a number of years that market's already here. They are travelling, they are travelling in greater numbers [situation awareness and adaptive capacity] … I think because we work in this industry, you work with tourism, you work with people … we bounced back extremely quickly [psychological resilience]. By our nature, we are [resilient] … We've got good ability to change [adaptive capacity], we are just meeting that [new] market [situation monitoring and reporting], and quickly establishing ourselves [adaptive capacity] … 90% of our clients now are Chinese nationals, we've adjusted to that. (Participant 11)

Drawing from the data, it becomes clear that quickly bouncing back psychologically and proactively searching for new opportunities to upgrade the business could potentially result in an enhanced situation monitoring for the business, which contributes to the resilience of their business. Also, as the second quote demonstrates, because of the resilience of the entrepreneur, they quickly bounced back and adapted the business to the newly recognized market opportunity. Previous studies have shown that resilient entrepreneurs are willing to adapt to changes and take advantage of new situations (Ayala & Manzano, 2014).

Conclusion

This chapter partially fulfils the call for more studies on how tourism SMEs cope and respond to disasters (Orchiston, 2012; Hall *et al.*, 2018). Psychological resilience requires the ability of people to transcend the equilibrium (Luthans, 2002). This chapter shows that through self-efficacy, behavioural characteristics and social capital, tourism entrepreneurs are able to bounce back and this positively impacts the business. As such, the chapter illustrates how psychological resilience is a resource that can be used by tourism entrepreneurs in times of adversity to build organizational resilience. In line with Korber and McNaughton's (2018) suggestion we found that entrepreneurial resilience is a term conflated by notions of individual or psychological resilience and organizational resilience. However, our findings clearly pinpoint to psychological resilience being a resource that activates innovation, creativity, situation awareness, monitoring and the desire to change the business model, among others that are inherent drivers of organizational resilience. Hence, we argue that psychological resilience can directly impact the organizational resilience of tourism enterprises, unlike the study of Prayag *et al.* (2020) which showed only indirect impacts.

These themes highlight how tourism entrepreneurs find their own way to deal with the aftermath of disasters, including the consequences they have on tourism businesses. We concur with the view that resilience can be developed, managed and enhanced as suggested in previous studies (Luthans *et al.*, 2007; Masten *et al.*, 2009). For tourism businesses to thrive post-disaster, local government and small business associations should offer resilience training programmes along with business development opportunities so that entrepreneurs can move quickly from a negative to an entrepreneurial mindset. As suggested in previous studies (Biggs *et al.*, 2012; de Vries & Hamilton, 2016; Orchiston *et al.*, 2016), a disaster offers opportunities to reinvent the business by changing business models and pursuing new business opportunities emerging from the changed reality.

Similar to the study of Cioccio and Michael (2007) we found that accumulated experience helped with business recovery, but we extend this finding by showing that previous experience activates psychological resilience which allows entrepreneurs to quickly respond to negative impacts on their businesses. Pre-disaster networks and social capital emanating from family and friends are important (Bullough & Renko, 2013). Hence, similar to the study of Biggs *et al.* (2012) we found social capital as an important driver of business recovery in small firms. However, we suggest that social capital enables small tourism firm owner-operators to use these trust networks to boost their own optimism, confidence and entrepreneurial mind set. This allows them to use psychological resilience as resource for achieving positive business outcomes. While this study focuses on entrepreneurs that describe themselves as resilient, there is a dire need to research those entrepreneurs that perceived themselves as not resilient and/or those who experienced business closure post-disaster to fully understand the nexus of psychological resilience and organizational resilience.

References

Aldrich, D.P. (2012) *Building Resilience: Social Capital in Post-disaster Recovery*. Chicago: The University of Chicago Press.

Aroian, K.J. and Norris, A.E. (2000) Resilience, stress, and depression among Russian immigrants to Israel. *Western Journal of Nursing Research* 22 (1), 54–67.

Avey, J.B., Wernsing, T.S. and Luthans, F. (2008) Can positive employees help positive organizational change? Impact of psychological capital and emotions on relevant attitudes and behaviors. *The Journal of Applied Behavioral Science* 44 (1), 48–70.

Ayala, J.C. and Manzano, G. (2014) The resilience of the entrepreneur. Influence on the success of the business. A longitudinal analysis. *Journal of Economic Psychology* 42, 126–135.

Bayer, K. (2017) *Quake-damaged SH1 north of Kaikoura to finally reopen after 'unprecedented' 13 month repair job* [online]. See http://www.nzherald.co.nz/nz/news/article.cfm?c_id=1&objectid=11959817.

Biggs, D., Hall, C.M. and Stoeckl, N. (2012) The resilience of formal and informal tourism enterprises to disasters: Reef tourism in Phuket, Thailand. *Journal of Sustainable Tourism* 20 (5), 645–665.

Braun, V. and Clarke, V. (2006) Using thematic analysis in psychology. *Qualitative Research in Psychology* 3 (2), 77–101.

Bullough, A. and Renko, M. (2013) Entrepreneurial resilience during challenging times. *Business Horizons* 56 (3), 343–350.

Christiansen, J., Christiansen, J.L. and Howard, M. (1997) Using protective factors to enhance resilience and school success for at-risk students. *Intervention in School and Clinic* 33 (2), 86–89.

Cioccio, L. and Michael, E.J. (2007) Hazard or disaster: Tourism management for the inevitable in Northeast Victoria. *Tourism Management* 28 (1), 1–11.

Compas, B.E., Hinden, B.R. and Gerhardt, C.A. (1995) Adolescent development: Pathways and processes of risk and resilience. *Annual Review of Psychology* 46 (1), 265–293.

Cooper, C., Flint-Taylor, J. and Pearn, M. (2013) *Building Resilience for Success: A Resource for Managers and Organizations*. Basingstoke: Palgrave Macmillan.

Cró, S. and Martins, A.M. (2017) Structural breaks in international tourism demand: Are they caused by crises or disasters? *Tourism Management* 63, 3–9.

Dahles, H. and Susilowati, T.P. (2015) Business resilience in times of growth and crisis. *Annals of Tourism Research* 51, 34–50.

Danes, S.M. (2013) Entrepreneurship success: 'The Lone Ranger' versus 'It Takes a Village' approach? *Entrepreneurship Research Journal* 3 (3), 277–286.

de Vries, H. and Hamilton, R. (2016) Why stay? The resilience of small firms in Christchurch and their owners. In C.M. Hall, S. Malinen, R. Vosslamber and R. Wordsworth (eds) *Business and Post-disaster Management: Business, Organisational and Consumer Resilience and the Christchurch Earthquakes* (pp. 22–34). Oxon and New York: Routledge.

de Vries, H. and Shields, M. (2006) Towards a theory of entrepreneurial resilience: A case study analysis of New Zealand SME owner operators. *New Zealand Journal of Applied Business Research* 5 (1), 33–43.

Duchek, S. (2018) Entrepreneurial resilience: A biographical analysis of successful entrepreneurs. *International Entrepreneurship and Management Journal* 14 (2), 429–455.

Fan, F., Zhang, Y., Yang, Y., Mo, L. and Liu, X. (2011) Symptoms of posttraumatic stress disorder, depression, and anxiety among adolescents following the 2008 Wenchuan earthquake in China. *Journal of Traumatic Stress* 24 (1), 44–53.

Ferraro, F.R. (2003) Psychological resilience in older adults following the 1997 flood. *Clinical Gerontologist* 26 (3–4), 139–143.

Hall, C.M., Malinen, S., Vosslamber, R. and Wordsworth, R. (2016a) Introduction: The Business, organizational and destination impacts of natural disasters - the Christchurch earthquakes 2010–2011. In C.M. Hall, S. Malinen, R. Vosslamber and R. Wordsworth (eds) *Business and Post-disaster Management: Business, Organizational and Consumer Resilience and the Christchurch Earthquakes* (pp. 3–22). Abingdon: Routledge.

Hall, C.M., Malinen, S., Vosslamber, R. and Wordsworth, R. (eds) (2016b) *Business and Post-disaster Management: Business, Organizational and Consumer Resilience and the Christchurch Earthquakes*. Abingdon: Routledge.

Hall, C.M., Prayag, G. and Amore, A. (2018) *Tourism and Resilience: Individual, Organisational and Destination Perspectives*. Bristol: Channel View Publications.

Jackson, D., Firtko, A. and Edenborough, M. (2007) Personal resilience as a strategy for surviving and thriving in the face of workplace adversity: A literature review. *Journal of Advanced Nursing* 60 (1), 1–9.

Jiang, Y., Ritchie, B.W. and Verreynne, M.L. (2019) Building tourism organizational resilience to crises and disasters: A dynamic capabilities view. *International Journal of Tourism Research* in press, 1–19.

Korber, S. and McNaughton, R.B. (2018) Resilience and entrepreneurship: A systematic literature review. *International Journal of Entrepreneurial Behavior & Research* 24 (7), 1129–1154.

Loh, J.M. and Dahesihsari, R. (2013) Resilience and economic empowerment: A qualitative investigation of entrepreneurial Indonesian women. *Journal of Enterprising Culture* 21 (1), 107–121.

Luthans, F. (2002) The need for and meaning of positive organizational behavior. *Journal of Organizational Behavior* 23 (6), 695–706.

Luthans, F., Youssef, C.M. and Avolio, B.J. (2007) *Psychological Capital: Developing the Human Competitive Edge*. Oxford: Oxford University Press.

Mann, C.L., Gillezeau, C.N., Massazza, A., Lyons, D.J., Tanaka, K., Yonekura, K., Sekine, H., Yanagisawa, R. and Katz, C.L. (2018) Fukushima triple disaster and the road to recovery: A qualitative exploration of resilience in internally displaced residents. *Psychiatric Quarterly* 89 (2), 383–397.

Masten, A.S. (2001) Ordinary magic: Resilience processes in development. *American Psychologist* 56 (3), 227–238.

Masten, A.S., Cutuli, J.J., Herbers, J.E. and Reed, M.J. (2009) Resilience in development. In C.R. Snyder and S.J. Lopez (eds) *The Oxford Handbook of Positive Psychology* (pp. 117–132). Oxford: Oxford University Press.

Mazzocchi, M. and Montini, A. (2001) Earthquake effects on tourism in central Italy. *Annals of Tourism Research* 28 (4), 1031–1046.

McDonald, G., Jackson, D., Vickers, M.H. and Wilkes, L. (2016) Surviving workplace adversity: A qualitative study of nurses and midwives and their strategies to increase personal resilience. *Journal of Nursing Management* 24 (1), 123–131.

McManus, S., Seville, E., Vargo, J. and Brunsdon, D. (2008) Facilitated process for improving organizational resilience. *Natural Hazards Review* 9 (2), 81–90.

Ministry of Business, Innovation & Employment (MBIE) (2017) *Kaikoura: Tourist Travel Behaviour and Recovery Framework*. Wellington: New Zealand Government.

Ministry of Economic Development (2011) *SMEs in New Zealand: Structure and Dynamics 2011* [pdf]. See http://www.mbie.govt.nz/info-services/business/business-growth-and-internationalisation/documents-image-library/Structure-and-Dynamics-2011.pdf.

Ministry of Transport (2017) *Economic Impact of the 2106 Kaikoura Earthquake*. Wellington: New Zealand Government.

Mzid, I., Khachlouf, N. and Soparnot, R. (2019) How does family capital influence the resilience of family firms? *Journal of International Entrepreneurship* 17 (2), 249–277.

Orchiston, C. (2012) Seismic risk scenario planning and sustainable tourism management: Christchurch and the alpine fault zone, South Island, New Zealand. *Journal of Sustainable Tourism* 20 (1), 59–79.

Orchiston, C. (2013) Tourism business preparedness, resilience and disaster planning in a region of high seismic risk: The case of the Southern Alps, New Zealand. *Current Issues in Tourism* 16 (5), 477–494.

Orchiston, C., Prayag, G. and Brown, C. (2016) Organizational resilience in the tourism sector. *Annals of Tourism Research* 56, 145–148.

Ozanne, L. and Ozanne, J.L. (2016) How alternative consumer markets can build community resiliency. *European Journal of Marketing* 50 (3/4), 330–357.

Prayag, G. (2018) Symbiotic relationship or not? Understanding resilience and crisis management in tourism. *Tourism Management Perspectives* 25, 133–135.

Ayala, J.C. and Manzano, G. (2014) The resilience of the entrepreneur. Influence on the success of the business. A longitudinal analysis. *Journal of Economic Psychology* 42, 126–135.
Bayer, K. (2017) *Quake-damaged SH1 north of Kaikoura to finally reopen after 'unprecedented' 13 month repair job* [online]. See http://www.nzherald.co.nz/nz/news/article.cfm?c_id=1&objectid=11959817.
Biggs, D., Hall, C.M. and Stoeckl, N. (2012) The resilience of formal and informal tourism enterprises to disasters: Reef tourism in Phuket, Thailand. *Journal of Sustainable Tourism* 20 (5), 645–665.
Braun, V. and Clarke, V. (2006) Using thematic analysis in psychology. *Qualitative Research in Psychology* 3 (2), 77–101.
Bullough, A. and Renko, M. (2013) Entrepreneurial resilience during challenging times. *Business Horizons* 56 (3), 343–350.
Christiansen, J., Christiansen, J.L. and Howard, M. (1997) Using protective factors to enhance resilience and school success for at-risk students. *Intervention in School and Clinic* 33 (2), 86–89.
Cioccio, L. and Michael, E.J. (2007) Hazard or disaster: Tourism management for the inevitable in Northeast Victoria. *Tourism Management* 28 (1), 1–11.
Compas, B.E., Hinden, B.R. and Gerhardt, C.A. (1995) Adolescent development: Pathways and processes of risk and resilience. *Annual Review of Psychology* 46 (1), 265–293.
Cooper, C., Flint-Taylor, J. and Pearn, M. (2013) *Building Resilience for Success: A Resource for Managers and Organizations*. Basingstoke: Palgrave Macmillan.
Cró, S. and Martins, A.M. (2017) Structural breaks in international tourism demand: Are they caused by crises or disasters? *Tourism Management* 63, 3–9.
Dahles, H. and Susilowati, T.P. (2015) Business resilience in times of growth and crisis. *Annals of Tourism Research* 51, 34–50.
Danes, S.M. (2013) Entrepreneurship success: 'The Lone Ranger' versus 'It Takes a Village' approach? *Entrepreneurship Research Journal* 3 (3), 277–286.
de Vries, H. and Hamilton, R. (2016) Why stay? The resilience of small firms in Christchurch and their owners. In C.M. Hall, S. Malinen, R. Vosslamber and R. Wordsworth (eds) *Business and Post-disaster Management: Business, Organisational and Consumer Resilience and the Christchurch Earthquakes* (pp. 22–34). Oxon and New York: Routledge.
de Vries, H. and Shields, M. (2006) Towards a theory of entrepreneurial resilience: A case study analysis of New Zealand SME owner operators. *New Zealand Journal of Applied Business Research* 5 (1), 33–43.
Duchek, S. (2018) Entrepreneurial resilience: A biographical analysis of successful entrepreneurs. *International Entrepreneurship and Management Journal* 14 (2), 429–455.
Fan, F., Zhang, Y., Yang, Y., Mo, L. and Liu, X. (2011) Symptoms of posttraumatic stress disorder, depression, and anxiety among adolescents following the 2008 Wenchuan earthquake in China. *Journal of Traumatic Stress* 24 (1), 44–53.
Ferraro, F.R. (2003) Psychological resilience in older adults following the 1997 flood. *Clinical Gerontologist* 26 (3–4), 139–143.
Hall, C.M., Malinen, S., Vosslamber, R. and Wordsworth, R. (2016a) Introduction: The Business, organizational and destination impacts of natural disasters - the Christchurch earthquakes 2010–2011. In C.M. Hall, S. Malinen, R. Vosslamber and R. Wordsworth (eds) *Business and Post-disaster Management: Business, Organizational and Consumer Resilience and the Christchurch Earthquakes* (pp. 3–22). Abingdon: Routledge.
Hall, C.M., Malinen, S., Vosslamber, R. and Wordsworth, R. (eds) (2016b) *Business and Post-disaster Management: Business, Organizational and Consumer Resilience and the Christchurch Earthquakes*. Abingdon: Routledge.
Hall, C.M., Prayag, G. and Amore, A. (2018) *Tourism and Resilience: Individual, Organisational and Destination Perspectives*. Bristol: Channel View Publications.

Jackson, D., Firtko, A. and Edenborough, M. (2007) Personal resilience as a strategy for surviving and thriving in the face of workplace adversity: A literature review. *Journal of Advanced Nursing* 60 (1), 1–9.

Jiang, Y., Ritchie, B.W. and Verreynne, M.L. (2019) Building tourism organizational resilience to crises and disasters: A dynamic capabilities view. *International Journal of Tourism Research* in press, 1–19.

Korber, S. and McNaughton, R.B. (2018) Resilience and entrepreneurship: A systematic literature review. *International Journal of Entrepreneurial Behavior & Research* 24 (7), 1129–1154.

Loh, J.M. and Dahesihsari, R. (2013) Resilience and economic empowerment: A qualitative investigation of entrepreneurial Indonesian women. *Journal of Enterprising Culture* 21 (1), 107–121.

Luthans, F. (2002) The need for and meaning of positive organizational behavior. *Journal of Organizational Behavior* 23 (6), 695–706.

Luthans, F., Youssef, C.M. and Avolio, B.J. (2007) *Psychological Capital: Developing the Human Competitive Edge*. Oxford: Oxford University Press.

Mann, C.L., Gillezeau, C.N., Massazza, A., Lyons, D.J., Tanaka, K., Yonekura, K., Sekine, H., Yanagisawa, R. and Katz, C.L. (2018) Fukushima triple disaster and the road to recovery: A qualitative exploration of resilience in internally displaced residents. *Psychiatric Quarterly* 89 (2), 383–397.

Masten, A.S. (2001) Ordinary magic: Resilience processes in development. *American Psychologist* 56 (3), 227–238.

Masten, A.S., Cutuli, J.J., Herbers, J.E. and Reed, M.J. (2009) Resilience in development. In C.R. Snyder and S.J. Lopez (eds) *The Oxford Handbook of Positive Psychology* (pp. 117–132). Oxford: Oxford University Press.

Mazzocchi, M. and Montini, A. (2001) Earthquake effects on tourism in central Italy. *Annals of Tourism Research* 28 (4), 1031–1046.

McDonald, G., Jackson, D., Vickers, M.H. and Wilkes, L. (2016) Surviving workplace adversity: A qualitative study of nurses and midwives and their strategies to increase personal resilience. *Journal of Nursing Management* 24 (1), 123–131.

McManus, S., Seville, E., Vargo, J. and Brunsdon, D. (2008) Facilitated process for improving organizational resilience. *Natural Hazards Review* 9 (2), 81–90.

Ministry of Business, Innovation & Employment (MBIE) (2017) *Kaikoura: Tourist Travel Behaviour and Recovery Framework*. Wellington: New Zealand Government.

Ministry of Economic Development (2011) *SMEs in New Zealand: Structure and Dynamics 2011* [pdf]. See http://www.mbie.govt.nz/info-services/business/business-growth-and-internationalisation/documents-image-library/Structure-and-Dynamics-2011.pdf.

Ministry of Transport (2017) *Economic Impact of the 2106 Kaikoura Earthquake*. Wellington: New Zealand Government.

Mzid, I., Khachlouf, N. and Soparnot, R. (2019) How does family capital influence the resilience of family firms? *Journal of International Entrepreneurship* 17 (2), 249–277.

Orchiston, C. (2012) Seismic risk scenario planning and sustainable tourism management: Christchurch and the alpine fault zone, South Island, New Zealand. *Journal of Sustainable Tourism* 20 (1), 59–79.

Orchiston, C. (2013) Tourism business preparedness, resilience and disaster planning in a region of high seismic risk: The case of the Southern Alps, New Zealand. *Current Issues in Tourism* 16 (5), 477–494.

Orchiston, C., Prayag, G. and Brown, C. (2016) Organizational resilience in the tourism sector. *Annals of Tourism Research* 56, 145–148.

Ozanne, L. and Ozanne, J.L. (2016) How alternative consumer markets can build community resiliency. *European Journal of Marketing* 50 (3/4), 330–357.

Prayag, G. (2018) Symbiotic relationship or not? Understanding resilience and crisis management in tourism. *Tourism Management Perspectives* 25, 133–135.

Prayag, G., Fieger, P. and Rice, J. (2019) Tourism expenditure in post-earthquake Christchurch, New Zealand. *Anatolia* 30 (1), 47–60.

Prayag, G., Spector, S., Orchiston, C. and Chowdhury, M. (2020) Psychological resilience, organizational resilience and life satisfaction in tourism firms: Insights from the Canterbury earthquakes. *Current Issues in Tourism* in press, 1–18.

Richardson, G.E. (2002) The metatheory of resilience and resiliency. *Journal of Clinical Psychology* 58 (3), 307–321.

Satake, K. (2014) Advances in earthquake and tsunami sciences and disaster risk reduction since the 2004 Indian Ocean tsunami. *Geoscience Letters* 1 (1), 1–13.

Shepherd, D.A., Williams, T.A. and Patzelt, H. (2015) Thinking about entrepreneurial decision making: Review and research agenda. *Journal of Management* 41 (1), 11–46.

Smith, R., McIntosh, V.V., Carter, J.D., Colhoun, H., Jordan, J., Carter, F.A. and Bell, C.J. (2017) In some strange way, trouble is good for people. Post-traumatic growth following the Canterbury earthquake sequence. *Australasian Journal of Disaster and Trauma Studies* 21 (1), 31–42.

Statistics New Zealand (2017) *Kaikoura Recovery by the Numbers* [online]. See http://archive.stats.govt.nz/browse_for_stats/industry_sectors/RetailTrade/kaikourarecovery.aspx.

Tierney, K.J. (1997) Business impacts of the Northridge earthquake. *Journal of Contingencies and Crisis Management* 5 (2), 87–97.

Tongco, M.D.C. (2007) Purposive sampling as a tool for informant selection. *Ethnobotany Research and Applications* 5, 147–158.

Tsai, C. and Chen, C. (2010) An earthquake disaster management mechanism based on risk assessment information for the tourism industry-a case study from the island of Taiwan. *Tourism Management* 31 (4), 470–481.

Youssef, C.M. and Luthans, F. (2007) Positive organizational behavior in the workplace: The impact of hope, optimism, and resilience. *Journal of Management* 33 (5), 774–800.

Wang, Y. (2009) The impact of crisis events and macroeconomic activity on Taiwan's international inbound tourism demand. *Tourism Management* 30 (1), 75–82.

Wang, Y., Tsai, C.H., Tsai, F.S., Huang, W. and de la Cruz, S. (2018) Antecedent and consequences of psychological capital of entrepreneurs. *Sustainability* 10 (10), 3717.

10 Ghost Towns and Tourism: L'Aquila, Italy Post-Earthquake

Daniel Wright

Introduction

Ghosts and the paranormal have a strong place in society. How we engage with them also varies. Ghostly experiences can be of a serious nature, locations where actual death and tragedy have occurred, a natural or man-made disaster zone, leaving a city (partially) abandoned. More light-hearted ghostly experiences also exist, such as haunted hotel experiences, ghost tours and places of entertainment, such as theme parks. Entertainment, such as TV shows and movies also offer a mixture of graphic and light-hearted visual exposures to ghosts and the paranormal and could arguably be playing a role in fuelling the demand for such experiences. Be it from a supply or demand perspective, ghost tourism exists, people have a fascination and interest with the supernatural. But to date, limited research has attempted to explore ghost tourism from the perspective of a local community and tourists, especially in a post-disaster context. In this case L'Aquila, a city in central Italy that was devastated by an earthquake in 2009 and which has in theory and practice become a ghost town – ghostly, vacant, unnerving and silent. In the context of the present chapter, as well as in comparison with the potential of other post-disaster tourism initiatives raised in this book, the study considered, what constitutes a place becoming a ghost town? Does a ghost town need to be empty, disserted or can it still retain signs of life, with residents living and working in the town (city)? If so, what do locals (still presiding in the city of having vacated) feel towards the city becoming and or being labelled as a ghost town? What do locals think of tourists visiting the city for the purposes of ghostly experiences? And once a place becomes or at least is given the label of a ghost town, do people visit because of such a label and do tourists see themselves as ghost tourists? Finally, can promoting an earthquake affected destination as a ghost town be of benefit to locals and or tourists? Through empirical research, this chapter offers original ideas and findings relevant to a novel approach to post-disaster tourism from the perspective of the local Aquilano community and tourists visiting the city.

Growth and Fascination of Ghosts and the Paranormal

Replace fear of the unknown with curiosity.

Interest in ghosts can often be attributed to discussions surrounding people's fascination with the paranormal, thus, this section will initially consider human fascination in paranormal activity and then consider the relationship and impact this can have on tourism engagement. According to Watt and Wiseman (2009: viii) 'paranormal beliefs are widely held in the population: around the world surveys consistently show that about fifty per cent of people hold one or more paranormal beliefs, and of these, about fifty percent believe they have had genuinely paranormal experience. Regardless of whether these beliefs and experiences are "correct", they are clearly an important part of what it is to be human'. Different definitions of the paranormal exist within the research, indicating an issue with the clarity and agreed understanding of such phenomena.

The Parapsychological Association applies the term *psi* (which includes experiences such as telepathy, clairvoyance, remote viewing, psychokinesis, psychic healing and precognition) in reference to experiences that challenge contemporary conceptions of human nature and that of the physical world and 'appear to involve the transfer of information and influence of the physical systems independently of time and space, via mechanisms we cannot currently explain' (Hill, 2011: 4). Irwin (2009) suggests that paranormal beliefs are phenomena that are scientifically inexplicable. Further noting, that such beliefs are 'generated within the non scientific community and extensively endorsed by people who might normally be expected by their society to be capable of rational thought and reality testing' (Irwin, 2009: 16). While Goode (1999: 16) defines the paranormal as 'the non scientific approach to a scientifically implausible event believed to be literally true'. From the definitions, there is a clear recognition, that the paranormal is non-standard to experienced reality. While people understand their experience and those of others within the dominant narrative of selfhood and Western culture, paranormal ideas, writings and practices offer different narratives relating to death and the afterlife and metaphysical matters (Hill, 2011). Such paranormal and inconsistent happenings are explained through culture and subcultures, where ideologies and beliefs are attributed meaning and explanation (Cardena *et al.*, 2000). It is not necessary for this study to explore the drives behind peoples' supernatural beliefs across the wide-ranging cultures that occupy our Earth. Discussions are complex as one might say that, from a religious perspective, a belief in God is to believe in the supernatural. Popular culture can also be credited with much of the fascination in the paranormal and, consequently, associations with ghostly experiences. Willingly or unwillingly, people are consuming and engaging with real and, more so, commodified death and suffering through audiovisual media channels, and popular culture is a key source (Sharpley & Stone, 2009). Our screens are awash with spooky and out of this world visual images, more so than at any other time in history. Television shows and movies have bought into the paranormal and supernatural theme and, today, audiences can experience all forms including vampires (*True Blood*), ghosts (*American Horror Story*), zombies, (*The Walking Dead*) and extra-terrestrials (*Prometheus*). Consequently, our interests are influenced by popular culture (Wright, 2016) and this is one form in which peoples' supernatural desires can be enhanced in contemporary society.

Ghost tourism: Demand and supply perspective

> Ghost walks and ghost hunting and ghost tourism and ghost shows on the television all seem to be growing in popularity. (Taylor, 2011)

The commercial exploitation of ghosts is not a novel practice. Davies (2007) recognises a shift from a historical approach in which communities would try to rid themselves of ghostly spirits to one where the ghost is seen as desirable and not as an unwelcome guest. Consequently, fuelling the demand for commercial marketing of haunted places which were limited in number pre-20th century. According to Davies (2007), early historical records expose the regularity with which crowds would gather at places and sites of alleged haunting and the benefits of local businesses in reaping the rewards. Davis (2007: 63) draws on the example of Horace Walpole's observation of the famous Cock Lane ghost of 1762 which allowed 'all the taverns and alehouses in the neighbourhood [to] make fortunes'. Eighteenth and nineteenth-century upper-class fascination with Scotland as a source of ghosts and the supernatural wonderings has been explored by Inglis and Holmes (2003). The demand and drive of ghost tourists have witnessed limited coverage in the academic sphere, as such tourists were previously more likely to be labelled as dark tourists (which will be considered in more detail shortly). The UK has a popular fascination with haunted inns and pubs, offering owners and local tourism providers the opportunity to experience a night at their spooky establishments (Holloway, 2010). Ghost tourism is popular globally, and according to Holloway (2010: 619), there are at least three main elements to contemporary ghost tourism.

(1) Hotels seeking guests that claim to be haunted.
(2) Companies offering tourist the chance to go ghost hunting.

(3) Ghost tours or ghost walks, which involve organised walks around cities and towns during which one takes in the sites of alleged hauntings, listens to ghost stories and in some cases (such as Edinburgh) watches actors perform the narratives behind the spooky happenings.

There are many examples of haunted hotels around the globe; a quick internet search will bring up a range of articles. For example, the world's most haunted hotels by *The Telegraph* (2018) provides an insight into some of these. The Stanley Hotel, Colorado (inventor of America's Stanley Steamer automobiles, 1907), has seen much paranormal activity, such as the presence of Elisabeth Wilson, a former house keeper (died in 1911), who is said to be present in her ghostly form, assisting guests to store away belongings in room 217. Also present is the sound of children running and giggling down the halls. Chillingham Castle, Northumberland, England (a 12th century castle) has a gruesome history of executions in its torture chamber. It is said only the brave should book a stay in the 'Pink Room', where guests report seeing blue lights flashing, a spooky phenomenon that has become known as the 'Blue Boy'. Hotel Burchianti in Florence, Italy, is apparently full of ghosts. Guests have witnessed children skipping down the halls, a woman knitting in a chair and a maid doing her cleaning round in the early hours. Room number eight in The Russell Hotel in Sydney, Australia, is said to be haunted by the spirit of a sailor. The 12th century castle, Dragsholm Slot, in Denmark, is said to be haunted by the ghosts of two women and the Earl of Bothwell (who was allegedly imprisoned in the castle and died in the cellars in 1578). Examples of haunted houses are often shadowed by history and the presence of death in some form or another. The examples shown here also attain a level of sincerity, in the sense that these are locations that function as hotels, but have the haunted label attached to them because of the paranormal activity experienced by guests. However, the consumer market is also awash with premeditated experiences.

For tourists who actively want to hunt ghosts, in line with Holloway's (2010) second category, they can purchase such experiences through companies such as Dusk Till Dawn Events (2018), 'The Ghost Hunting Company'. The company offers tourists the chance to spend the night in a haunted building on frightening overnight ghost hunts at some of the UK's most haunted locations. In another corner of the globe, companies such as the Singapore Paranormal Investigators and the Asia Paranormal Investigators regularly take inquisitive groups and individuals on spooky hunts around the city-state (Hutt, 2015). In the UK, tourists can even purchase ghost hunting through Virgin Experience Days (2018). Consumers will 'join psychics and paranormal investigators on a night of séances, experiments and vigils you'll never forget'. The offer entices potential buyers with the following blurb

> Witness for yourself what supernatural beings lurk in the dark as you take part in a ghost hunt at one of a choice of locations. Whether you're a budding ghost hunter or are just looking to experience something strange and inexplicable, this experience is the perfect outing. Throughout the night you will try to detect their presence using the latest ghost hunting equipment and try to contact the spirits through means of dowsing and glass moving. (Virgin Experience Days, 2018)

In accordance with Holloway's (2010) third category, ghost tours, the top 10 tourist experiences in the world, as advertised by Tripadvisor (2018), at the time of writing, include Bill Spectre's Ghost Trails in Oxford, England. Harpers Ferry Ghost Tours in West Virginia, United States; McGee's Ghost Tours of Prague; Mysteries of Paris Vampire and Ghost Tours; Ghost Bus Tour Bus in Dublin, Ireland. In the United States, many metropolitan areas offer ghost tours, or a ghost investigation club, many of which are somewhat new, being established around 2007 onwards (Molle & Bader, 2014).

Clearly, there is some overlap between Holloway's (2010) three categories, as some haunted hotels (category 1) are used as locations for ghost hunting (category 2) and can also be used for ghost tours (category 3). Interestingly, what is available on the consumer market is a large variety of different experiences and types of locations and attractions. As noted, the relationship with death and the afterlife offers a certain fascination for people seeking paranormal activities. A clear link here for ghost tourism is its relationship with thanatourism and the more commonly applied term of dark tourism, which implies a relationship between travel to places associated with death, dying and disaster (Foley & Lennon, 1996; Lennon & Foley, 1999; Sharpley & Stone, 2009). In this context, Stone's (2006) dark tourism spectrum could also be of interest. In the spectrum, Stone discusses different shades of darkness (lighter to darker), in reference to the seriousness of the attraction. The levels of darkness and lightness here could be applied to the level of entertainment added to the experience on offer. Where higher levels of commercialisation and marketing are purposely aimed at targeting tourists for a ghostly experience, it could be suggested that they fall on the lighter end of the spectrum. Whereas places with lower levels of commercial activity and advertising could be placed at the darker end of spectrum. However, the fluid nature of places and attractions on the spectrum is evident. Places and attractions offer different experiences to tourists and simply considering Stone's spectrum allows one to consider the range and diversity of locations which have a relationship to death and suffering. For this study, while ghost tourism could be placed under the dark tourism umbrella term, the aim is to explore ghost tourism as an independent entity. However, Stone's spectrum is useful from a supply perspective in understanding destinations and their level of darkness, and will be considered in line with L'Aquila.

This chapter is interested with Holloway's (2010) third category, ghost tours or ghost walks, which is a type of ghost tourism that has become an international phenomenon and can be located around the globe in most major cities in Europe, the United States, Australia and New Zealand. However, the description Holloway (2010) offers includes a level of commercial activity, one with some form of tourism infrastructure of at least structure and purpose. But not all ghost towns offer such guided and proactive approaches to tourism. Many ghost towns do not offer tourism services and facilities. Instead, what they offer is just the visual spectacle of what remains of a former existence. This is very much the reality for L'Aquila, the location under examination here. Table 10.1 provides some examples of cities that are frequented by tourists due to their ghostly nature.

Limited academic research has been conducted on tourist motivations to visit ghost tourism attractions, and those that have been conducted often fall into the dark tourism scholarly field. Additionally, the ghostly attractions that have been researched are often locations where there is a higher level of tourism management and infrastructure. For example, research by Bissell (2009) explored tourist motivations for visiting ghost tours jointly operated by Heartland International Travel Tours and Muddy Water Tours. These tours were 2.5 hour long guided bus tours of Winnipeg, Canada, that commenced at the Winnipeg Via Rail Station and ended at the Hotel Fort Garry in downtown Winnipeg. The researchers' conclusions were that visitors were interested in the history recounted in the tours. Rittichainuwat (2011) explored differences between Asian and Western tourists and whether or not beliefs in ghosts could deter tourists from travelling to disaster-hit destinations. The study examined travel barriers associated with tsunami-hit destinations and assessed cultural differences in beliefs between Asians and Western tourists and also across different demographic variables. The research found that 'cultural beliefs and norms should be understood as essential considerations in crisis management and destination recovery so that marketers integrate such beliefs and norms in effectively planning marketing communication after a disaster' (Rittichainuwat, 2011: 425). Furthermore, the author highlights the physical and psychological damage left behind after a disaster. Consequently, there is a need to understand the psychological recovery, which can limit the speed at which a destination recovers. Importantly, for some cultures and people who hold strong beliefs in ghosts or have experienced trauma, a destination will continue to be a place of mourning unless all the corpses have been discovered and appropriately buried and or religious ceremonies carried

Table 10.1 Ghost tourism cities

Ghost city	Description
Craco, Italy	This hillside ghost down was founded in the 8th century and sits on a cliff that is 1312 feet off the ground. The city emptied due to various natural disasters. In 1963, many evacuated after a landslide; in 1972 a flood made conditions even more precarious; and in 1980 an earthquake caused the town to be abandoned in its entirety. A locked gate surrounds the city, so visitors must book a guided tour.
Terlingua, Texas, USA	When the Chisos Mining Company opened in the mid 1800s, workers and their families quickly relocated to Terlingua, Texas. The population was around 3000 at its peak in 1903, but, as of the last Census in 2010, only 58 people remain. Those who still live there reside in 'Terlingua Proper' and make good business off the frequent tourists who stop by to see the abandoned churches and buildings that still stand, as well as visitors to the surrounding Big Bend parks.
Pripyat, Ukraine	This city in northern Ukraine is probably the most famous ghost town in the world. It was home to almost 50,000 people before being evacuated in April 1986, when the Chernobyl Nuclear Station suffered a meltdown.
Hashima Island, Japan	Hashima Island was once known for its undersea coal mines, which began operations in 1881. The island hit peak population in 1959 with over 5000 residents (mine workers and their families), but once the mines started to run out in 1974, most people left. The once thriving island is now completely abandoned, with the exception of the sightseeing tours that drop off boatloads of tourists each day who come to see the abandoned homes, stores and streets.
Kolmanskop, Namibia	Kolmanskop was at its liveliest in the early 1900s, when German miners came to the area to hunt for diamonds. With them, they brought German architecture, giving the desert area an opulent, out of place look. The town featured a ballroom, a hospital and a bowling alley among other amenities. The town's decline began shortly after World War I, but the final nail in the coffin was the 1928 discovery of a diamond-rich area along the coast. Most of Kolmanskop's residents hurried to the new hotspot, leaving their belongings and the town behind. Kolmanskop has been slowly getting eaten by the desert ever since. Tours to Kolmanskop can be booked in the nearby coastal town of Lüderitz.
Ross Island, India	Vegetation has all but consumed the remains of the island, which was once referred to as the 'Paris of the East'. In its prime, the island in the Andaman Islands was home to British government officials, as well as a penal settlement set up after the Indian Rebellion of 1857. The British residents made it their home with extravagant dance halls, bakeries, clubs, pools and gardens, until 1941 brought an earthquake and an invasion by the Japanese. Ross Island was then alternately claimed by the Japanese and British until 1979, when the island was given to the Indian Navy, which established a small base there. Today, tour groups visit the island almost every day. In December 2018, the Indian government officially renamed the island as Netaji Subhash Chandra Bose Island as a tribute to Subhas Chandra Bose, an Indian nationalist.

Source: Adapted from Daire (2017).

out (Rittichainuwat, 2011). An interesting concept the author touches on is the short-term effect of 'ghost panic', which can exist in a disaster destination, and is often experienced only until a place becomes crowded. Thus, once businesses, social life and tourism pick up, the 'ghost panic' is less likely to be an issue for people with stronger beliefs. The physical reconstruction and the psychological progress are significant in reducing the 'ghost panic' syndrome. Thus, the physical rebuilding of a destination and the injection of ordinary day-to-day living, such as a social existence in the local community and a post-mourning state, respectively, are essential in moving a destination beyond the initial ghostly nature.

Defining ghost towns

So what is a ghost town? The internet is awash with ghost town articles. Often, the articles present towns and cities that are now abandoned. As one article notes, 'There is nothing more haunting than a once-thriving town that has been abandoned and consumed by the elements' (Miklós, 2013). In another article, Daire (2017) suggests, 'Nothing can quite intrigue and horrify us in equal parts like a town that was abandoned in its entirety'. Table 10.1 provides some inclination of the various aspects that could constitute a ghost town. Remains and ruins of some form; impacts of natural or manmade disasters; the evacuation of its residents; and the end of a working industry (such as mining) are all common themes. Baker (2003) suggests that ghost towns must have tangible remains for visitors to see. The physical aspects are clearly important, with ruins often a common element, deserted buildings and belongings left behind, all aspects of what might be present in a ghost town. Significantly, however, towns do not need to be entirely abandoned to be named a ghost town. In the book *Legends of America* Weiser (2010) offers a definition of ghost towns. Weiser (2010) describes them as any historical town or site that leaves evidence of a town's pervious glory. This can be witnessed through the closure of businesses, minimal services and scattered ruins and rubble. Importantly, Weiser (2010) notes that some locations categorised as ghost towns still have inhabitants, even if local inhabitants do not accept their town as being classified and or named in such a manner. However, Weiser (2010) stresses that historians and other academics will continue to refer to them as ghost towns if the reason or purpose for their former 'boom' or glory is no more. This point is further addressed by the author of several popular ghost town books, Philip Varney (1994). Varney (1994) defines a ghost town as 'any site that has had a markedly decreased population from its peak, a town whose initial reason for settlement (such as a mine or railroad) no longer keeps people in the community'. A ghost town, then, 'can be completely deserted … or it can be a town with genuine signs of vitality' (Varney, 1994: xiii–xiv).

The case study discussed here, L'Aquila, is located in Italy. Demetri (2012) offers some useful clarity on the use of the term 'ghost towns' in Italy and explores a range of examples of Italian ghost towns that have become abandoned, in full or part, due to earthquakes. Examples include the town of Bussana Vecchia in Liguria, Giardino di Ninfa in the region of Lazio, the Beneventan town of Tocco Caudio and the Sicilian ghost town of Poggioreale located in the province of Trapani. Due to earthquakes, the locals were often required to leave, forced to relocate due to safety issues, rebuilding requirements or a lack of amenities. Some of these towns to this day remain completely abandoned, some partially inhabited and others have gained a new lease of life and are gradually reforming. Demetri's (2012) article was entitled, 'Città morte: Ghost towns of Italy'. Translated, *città*, meaning *city* and *morte*, meaning dead, thus translated as dead city. According to Demetri, *Città morte* for Italians, are the equivalent as ghost towns for English speakers (and not necessarily dead cities). The literal translation of ghosts in Italian is *fantasma*. This recognition is important in order to avoid translation issues during primary data collection during interviews.

Case study: L'Aquila, Italy

> L'Aquila: the wounds inflicted by the lethal earthquake seem slow to heal. The city, fled on a deadly night of six years ago, still remain suspended in time, showing the signs of an abruptly interrupted life.
> (UNFrame, 2015)

The city of L'Aquila is located in central Italy. The city, previously a centre of commercial activity, with a high population of university students and providing services for tourism, was hit by an earthquake (measuring 5.8 on the Richter Scale and 6.3 on the Moment Magnitude Scale) on 6 April 2009 at 3.32am. The disaster left 309 people dead, 1500 people injured and around 70,000 homeless. The earthquake saw the destruction and damage of around 10,000 buildings (Eggleton, 2010; McLaughlin, 2010). It was evident soon after the earthquake that tourists began to visit the city (Wright & Sharpley, 2018). However, the city of L'Aquila, offered very little in terms of physical services and facilities for tourists when visiting and or any interpretative information which would allow tourists to further engage with and understand what happened in the city or what the city used to be like (Wright, 2014). Tourists were very much left to their own imaginations when visiting and, apart from a few posters, flyers and memorials, there was almost an anti-tourism approach post-earthquake, as locals felt invaded and exploited (Wright, 2014; Wright & Sharpley, 2018).

Applying Stone's (2006) dark tourism spectrum, L'Aquila would likely fall on the darker end of the spectrum, as it is a place of actual death and suffering, offering more educational opportunities (and not entertainment driven) and limited tourism infrastructure. Similarly, taking a dark tourism supply perspective, Sharpley (2005) recognised two types of dark tourism categories, 'Accidental Dark Tourism Supply', attractions and sites that have become tourist attractions unintentionally, and 'Purposeful Dark Tourism Supply', attractions and sites which have intentionally been developed and established in order to create/meet the demand of tourists seeking to satisfy their fascination. In this instance, L'Aquila would be seen as an accidental dark tourism supply location, as the earthquake unintentionally created a location in which tourists could encounter death and suffering.

Articles soon after the earthquake recognised the ghostly nature of the city of L'Aquila. For example, the title *'One year after the earthquake, L'Aquila remains a ghost town'* by DW Akademie (2010). An article by Sharp (2010) reported that the local people of L'Aquila felt that their mayor had neglected the city, ultimately tuning the centre into a ghost town. Similarly, Dinmore (2009) noted that L'Aquila, the medieval town, has been turned into a ghost city. Moreover, in the article the author suggests that L'Aquila has become nothing more than a monument to the disaster that destroyed it, particularly with respect to the issues surrounding the reconstruction of the city centre. More recently, an article by UNFrame (2015) entitled *'L'Aquila, ghost city for six years and on'*, explores the challenges faced by the locals in their attempt to rebuild the city, but how its remains a tragedy and a visual representation of the earthquake that ravaged the city. Agnew (2016) noted the following: 'to say that the earthquake still casts its dark shadow over L'Aquila is to state the obvious. The heart of the town, the centro storico (historic centre), today is still a ghost town, largely unrebuilt with the exception of a handful of buildings, most of them either public administration or churches'. Agnew (2016) asks the following question, 'So why is the centre still a ghost town?', and suggests that the answer can be attained by exploring the original decisions taken by the Silvio Berlusconi government in 2009, and 'the failure to immediately set about the reconstruction of the town centre created an uninhabited ghost town' (Agnew, 2016).

As noted above, ghost in Italian is translated as *fantasma*. Many articles also run with this in their headlines. One example is *La Repubblica* (2018), which provides visual images of the devastation, with a title *'L'Aquila, città fantasma'*. Again, in *La Repubblica*, an article by Spinelli (2012), exploring the challenges the city faced, leads with the headline: *'L'Aquila, la città che non c'è più invasa da New Town fantasma* [L'Aquila, the city that no longer exists invaded as a new ghost town]'. An article by Santilli (2013) runs the headline: *'L'Aquila: città fantasma che vuole rinascere* [L'Aquila: a ghost town that wants to be reborn]'. Puliafito (2011) went with the headline, *'Intorno a L'Aquila paesi fantasma e zone rosse* [Around the Eagle, ghost towns and red areas]' (the eagle is the cities symbol). More recently, ANSA (Agenzia Nazionale Stampa Associata) (2018) highlighted the continued challenges of the reconstruction and the living and social condition went with the headline: *'Il centro storico dell'Aquila ancora fantasma* [The historic center of L'Aquila still a ghost]'. The above articles clearly identify the ghostly nature and the application of the term 'ghost town' by the media and locals. L'Aquila, post-earthquake, became a ghost town and continues to be one (at the time of writing). These articles and many others are evidence of this, and consequently, it offers an opportune location in which to explore ghost tourism from the perspective of locals and tourists.

The research: Study methods

The research for this study was conducted in L'Aquila during the summer of 2015. Two distinct groups, local community members and tourists, were identified in order to obtain the necessary data to explore the research objectives. In order to obtain participants that fell under 'local community members', an approach similar to the snowball sampling method was taken. Snowball sampling is seen as a non-probability technique and sometimes referred to as accidental sampling (Babbie, 2011). Snowball sampling was valuable as it is a procedure often applied when members of a specific community are difficult to locate and snowball sampling is frequently used during exploratory research. It requires some selective consideration of each participant by the researcher once interaction commences, to ensure the participant is appropriate for the study. Locals were interviewed in different locations, including private homes and cafes. Due to the nature of the research and the state of the city (at the time of research), accessing tourists was a little more complex. Here, the researcher's approach was one commonly identified as convenience sampling, also known as opportunistic and availability sampling (Brotherton, 2015; Long, 2008; Mason, 2014). In order to obtain tourists for participation in the study, the researcher approached tourists on the streets of the city centre. In consideration of the participant's capability of communicating in either Italian or English, participants were asked if they would be willing to contribute to a study exploring tourism in the city surrounding the research themes. In both sampling circumstances, no prior inclusion criteria were considered or deemed necessary in the selection of participants, other than those of falling under the category of being a tourist or a local resident.

A total of 32 interviews were conducted, 16 with local residents and 16 with tourists. A semi-structured approach to interviews was taken, as this allowed the researcher to approach the interviews in a more flexible manner and, in so doing, ensuring participants were comfortable and at ease. Significantly, this allowed discussions to evolve naturally in line with the participants reflections and feelings towards the themes in question (Gillham, 2005). All the interviews lasted around 5–15 minutes. Participants were made aware of the nature of the topics, and provided their consent to participate, to be recorded and they allowed their views to be expressed in the research findings. Participants were informed that their anonymity would be maintained, other than the details shown in Tables 10.2 and 10.3. Participants

Table 10.2 Research participants: Local community members

Local participants	Nationality	Gender M/F	Age
1	Italian	M	18–25
2	Italian	M	36–45
3	Italian	F	56–65
4	Italian	F	36–45
5	Italian	F	46–55
6	Italian	M	66–75
7	Italian	M	56–65
8	Italian	F	25–35
9	Italian	M	46–55
10	Italian	F	66–75
11	Italian	M	25–35
12	Italian	M	46–55
13	Italian	M	25–35
14	Italian	F	56–65
15	Italian	F	76+
16	Italian	M	76+

Table 10.3 Research participants: Tourists

Tourist participants	Nationality	Gender M/F	Age
1	Italian	M	46–55
2	Italian	M	36–45
3	Italian	F	56–65
4	French	M	18–25
5	French	F	25–35
6	Italian	M	46–55
7	Italian	F	46–55
8	English	F	56–65
9	English	M	56–65
10	Albanian	M	25–35
11	Croatian	M	36–45
12	Italian	F	36–45
13	Italian	F	18–25
14	Italian	F	18–25
15	USA	F	66–75
16	USA	M	66–75

were asked what age bracket they were in rather than their specific age (age categories were 18–25, 25–35, 36–45, 46–55, 56–65, 66–75, 76+). Interviews were conducted in either Italian or English (and captured on a Dictaphone). Interviews were transcribed into English, with the appropriate translation. The transcriptions were then subjected to a thematic analysis in order to identify themes, patterns and subjects and issues relevant to the study (Braun & Clarke, 2006).

Research analysis and discussion

As identified in the introduction, the broad purpose of this research was to explore initially how local community members felt towards their city, their home, becoming and or at least being labelled as a ghost town. Furthermore, research set out to explore how tourists viewed L'Aquila and its ghostly nature. It was identified above that the city of L'Aquila can be viewed and experienced as a ghost town, and this was evidenced in the manner in which the city has been portrayed and referenced in various channels and outlets, such as the media, since the 2009 earthquake. Additionally, the term and definition of a ghost town were also considered and consequently, L'Aquila was categorised as a ghost town, or in Italian, *Citta Morte* or *Citta Fantasma*.

The research findings exposed some interesting themes and opinions, not only between the locals and tourists but also within the two groups of respondents. Clear patterns emerged between local residents who felt that the city was not dead and should not be termed a ghost city and those that thought it was. From a tourist perspective, findings identified that the majority of tourists were motivated to visit L'Aquila due to curiosity and an interest to see how the city had progressed since 2009. Interestingly, what tourists eventually realised is that, when in the city, they were having experiences that could be commonly associated with ghostly and paranormal experiences. As tourists reflected on feeling uneasy, sometimes scared and freighted at exploring the empty streets and gazing into the abandoned buildings. Tourists felt that the city did have the aura of being spooky and ghost like, but for many of them, this became more evident during the interviews as they reflected on their experiences. Both parties (locals and tourists) were asked if they thought promoting the city as a ghost town would be a popular and useful approach to attracting and engaging tourists. Here, there was a contrast in opinions; while locals felt this would be an inappropriate term, tourists felt that the locals could use this as an appeal to tourists to bring in more money, especially because so little has progressed in the city and the lack of services and facilities available to tourists. The following sections will now explore the responses and themes in more detail, presenting actual participant responses.

Local resident's responses to ghost town L'Aquila

Research first set out to explore local resident's attitudes towards L'Aquila being termed a ghost town or dead town. The researcher was curious to know if locals felt that their city resembled a ghost (fantasma) and or dead (morte) city. When posed with the question, does your city feel like a ghost town or dead town?, one respondent suggested the following:

> Fantasma (ghost), morte (dead) does it matter what name you use? I mean look around you, look at the state of the city. There is so little movement, so few people, business. This city has been like this for six years. So little activity, so little social life. Yes, it feels like a haunted place.

Similarly, another respondent also drew on the lack of social life and how this provides the feelings of a city that holds a ghostly nature.

> The heart and soul of this city was drained straight after the earthquake. Since then the social life has gone, yes, people try to recover it, but like before, no. So certainly, for people who visit, even the locals, this place feels like a ghost town, or a dead city, however you want to call it. All you have to do is walk the streets, look into the empty buildings and you feel like this place is dead.

However, seeing the city as dead is not necessarily that easy. And some participants were a little more cautious to term the city in such a manner, as can be seen in the following two responses:

> You can use these terms, I think either is powerful when trying to express emotion and to express the pain and frustration of what has happened here. So little has taken place in the city since the earthquake that people are annoyed. So using such terms can really emphasise the seriousness of what happened here, it captures people's attention. When I walk the city streets, I see the city in the past, what it used to be like, now over time, a new reality, new thoughts, new memories exist. So now, when I think of my city, the more recent memories are in my mind. These are not as nice, the feelings are not as nice. So to your question, yes, this city does have a sense of emptiness, closed, cold, and if you want, ghostly and even dead. But not everywhere, because there is still some movement here, there are still people and some businesses, construction work. But for someone who visits for the first time, then yes, these ideas of ghost and dead would be more true, they do not have past memories of our city.

> This is not a dead city, there are people working and living here, of course not like before but still, this is our home our life, I am not dead, these people are not dead. I do not like these words, I do not think they provide the right image for our city. We need more positive, encouraging discussions, to help the city move forward. Maybe for other people this might feel like a dead or ghost city, but not me.

Overall, the respondents showed frustration towards their city and the lack of development since the earthquake, and consequently understood why terms such as ghost and dead would be used. This was best put by the following:

> Many words have been used to describe this city over the years, from local people, visitors, the media, politicians, everyone who visits have their own ideas and ways of describing what they see. People use words like dead, ghost, scary, empty, horror, cold, forgotten, many, many words. What is important is the message they take with them. And that message is the reality of what is here, of what they see, what they experience when they visit. That message is this city has had a traumatic few years, a very difficult time for our people. But what they see, a lot of is empty buildings, empty streets, not much social life, it's improving, but very slowly. They see the real L'Aquila, and yes, for many is can be unsettling.

What became apparent through the interviews with locals was that their perspective of the city was different from that of people who had never previously visited. As captured above, the locals have memories before the earthquake, what the city used to be like, where they once socialised with friends, went out for meals, where they would go to work. This memory bank is clearly important when considering the current state of the city and when confronted with the new reality. But, as noted, over time, that reality begins to change, as the more recent memories become stronger and ever present in the minds of the locals. So, even for locals, the memories of the past can eventually be more like those experienced by tourists, as the current L'Aquila becomes their reality, as observed here by the following participant:

> It is hard for us, this city had so much life. Then in a few seconds, everything changed. Yes, the city looks like a ghost town, one you see in the movies. I think it is even more strange for visitors who have not visited L'Aquila before, we are up in the mountains, a big city, but a little detached from anywhere else, so you come here and you see this big city and its current state and you think, wow, what happened here, where did everyone go? It can be a little uneasy to look at. It's a strange feeling walking around such a big place which has so little human activity. When I walk around, I still think of my past life here, but slowly, this past fades away, and the real L'Aquila, the one I experience now, is the one all around me.

This participant response truly captured the essence of the city:

> During the day you walk through the streets, silence, little noise, not many people, the buildings are empty, broken, forgotten, they remind us of the life we once had, as they remain stuck in the past. But this is

during the day, have you walked through these silent streets at night, there you have your ghost town, your dead city, it can be a very frightening experience, especially if you are new to this place.

Locals were also asked what they believed motivated tourists to visit the city. There was a lot of similarity in the responses, as locals drew on the tourists desire to see the damage caused by the earthquake and the level of work, or lack of construction, that has taken place since. Some locals commented on some of the memorials that exist in the city in memory of the victims, such as the well-known student house (Wright & Sharpley, 2018). There were also comments reflecting the interest of tourists in visiting L'Aquila as it continues to be a beautiful location in the mountains, scenic and offering fresh air, revitalising, even amongst the ruins. The following comments best capture these sentiments:

> A lot has been said about our city in the news, the media, what happened in L'Aquila, what has happened since the earthquake, what hasn't happened, lots of lies, we have become a forgotten and embarrassed city for the government. This is why tourists come to the city, they want to see the truth for themselves. Only visiting here really allows one to witness the reality of what has happened here.

Another respondent commented:

> I do not know why people want to visit this city, I would not want to go and see other people's tragedy. But L'Aquila is an interesting place, especially for the people of Abruzzo. A place that is unique to see. People would come here to get away from the sea, the enjoy the beautiful mountain air, to breath and recharge their batteries. And this still exists, the air is still pure here. But now, there is another tourism, this interest in the disaster, the damage it caused. Otherwise there is no much too see, just ruins, maybe the student house, this was publicised a lot in the media. Our iconic churches have been damaged, some badly destroyed.

Overall, the data attained from the locals suggested that, whilst they believed the majority of tourists visited to see the damage and post-disaster development, they also understood why the city was labelled a ghost town and that tourists could encounter ghostly experiences. More so, the locals felt that their experiences of the city were different to tourists due to the past memories (pre-earthquake) that they could draw on. However, over time, even their view of the city was changing as, in time, the current state (post-earthquake) becomes their new reality. But what about tourists? Tourists who have never visited the city before are not able to capture the past, they do not have feelings, memories and emotions to draw on. All they have is what is in front of them, what is visible to the eye. The physical reality that they are confronted with when visiting the city.

Tourist responses to ghost town L'Aquila

Throughout the street interviews, the focus was on tourists' opinions and experiences whilst in L'Aquila. The aim was to explore if they felt they were visiting a ghost town and if they saw themselves as ghost tourists. In general, tourists, when asked about the ghostly and dead nature of L'Aquila, felt that the city certainly captured some of the essence of what a ghost town should be, the idea of the place being abandoned, lacking a social and working soul. The overwhelming physical destruction and the incredible silence that is present was often what tourists reflected on. For example, in reference to the physical ambience and its ghostly nature the following participant responded with the following:

> It is a lot to take in, I mean, look around us, there are so many empty buildings, the entire city, all around you, everywhere you walk, down the main streets, the side streets, you can look through the doors, some are open, some are blocked, this place looks like it has been abandoned. It feels strange, nothing I have experienced before. I do think it feels like a ghost town, but more so now that you have mentioned it.

One respondent did not feel that the term dead was an appropriate reference, but did associate with the ghostly nature:

> I do not think this is a dead city, there is movement, reconstruction, a few bars, some people, so I'm not sure if dead it the correct word. But ghostly, yes, this does capture the feeling I have experienced, when I look into some houses and buildings, I feel a little frightened, I think that is normal. Empty buildings and deserted streets, it is unsettling.

It was common for respondents to be a little unsure if the term ghost town was suitable because, while the city is so destroyed, it is trying to rebuild itself, and this means that there is some movement and activity. This was best captured by the following participant's comment:

> I have heard much about L'Aquila in the news, so I was interested to visit and see for myself. Many articles have called this place a citta morte or fantasma. You can see why they say this, so many empty buildings, but the city is trying to give new life, and for me this is important, it is not a disserted ghost town with no life. Yes, large sections are abandoned, but the motivation is to continue.

Clearly, there is some confusion between what people might perceive or define a ghost town. As noted above, Weiser (2010) and Varney (1994) have suggested that a ghost town can show signs of life. On occasions, participants described their experience as something that they have witnessed on TV and in the movies, as noted here:

> It feels like something on the TV, in the movies, its incredible, I would recommend anyone to come and see this place. Yes, it is emotional as well, even if I have no reference to places, I still have emotions, thoughts, ideas, I can envisage scenarios in my mind, because I know what ghostly places are like. Because I have seen movies and this is similar.

Whilst the physical appearance of the city was clearly overwhelming for tourists visiting for the first time, many respondents did refer to the non-physical elements. Here, reference was given to the deeper creative imagination:

> A lot happened here, a lot of suffering and pain, people died here, in these buildings ... but I do not know which ones, this makes it even more creepy, because when you look into the buildings, you do not know what you are looking into, at first you stare at what you can physically see, brick work, broken concrete, wooden frames, old doors, but there is little in these places to see, but they do give you a sense of unease. But then you see a house with some old furniture, maybe some clothes or toys, pans, pots, plates, then my thoughts change. I think less of the physical things I can see, and I start to think of the people, the victims, this is a different feeling, more scary, more personal. Then I think the ghostly feelings can come through.

One of the tourists spent an evening walking around the city to capture photos at night. This experience, which was highlighted by a local earlier, is said to be even more ghostly. This is the tourist's response to that evening experience:

> I wanted to take some photos in the night, I thought that the place would be even more incredible, my wife did not want to join me, she was too scared, ha. But when I walked the streets at night, it was about two, three in the morning, this place became even more scary. It was so silent, it was a really haunting place, the buildings look even darker, more mysterious, disturbing. The lighting at night is different, the shadows created, captivating. I must admit, even I felt a little scared at times. It was an amazing experience, I had the whole city to walk around and no people, I do not think you can experience this in many places, not on this scale.

From the interviews, it was clear that tourists could recognise the ghostly nature of the city. While it was not always their original thought, when questioned on it, it became more apparent to them. With this in mind, another question was, do they see themselves as ghost tourists?

The answers were overwhelmingly against this notion. The tourists often stated that they were curious to see the destruction of the city, the damage, because they had heard a lot about the city, but they did not see themselves as ghost tourists. The following responses best capture this.

> Me, a ghost tourist, no, I don't think so. I have come here to see the destruction, the damage, but not ghosts.

> I would not call myself a ghost tourist, but then I do not know what type of tourist I would be, a disaster tourist maybe, but me, I am just curious to see what has happened here. You hear lots on the news, in papers, from other people, I was just curious to visit the city. To see what is happening with my own eyes.

However, while many tourists did not see themselves as ghost tourists, on reflection, many began to appreciate that a large part of their experience could actually be associated with feelings of a ghostly nature. The following responses best capture this:

> I came here to see the damage, what has happened in the city in the past 6 years. But the feeling I have had since being here is very strange. I did not expect to feel so uneasy walking around, especially down the smaller streets and peeking into empty buildings. My friend dared me to go in, I said no way, I was a little scared. It's strange, having this discussion with you has actually made me realise that there is an element of feeling frightened when experiencing this city, not something I expected before visiting.

> I do not think you can be prepared for this experience, it's not what I expected, I am not sure what I expected to see or feel. But I have seen great damage, a lot of empty and scary looking buildings, especially when you look inside them. Broken windows, vacant rooms, everything feels old and forgotten. This is not normal, not a normal experience, it is unsettling, and like you mentioned before, it's like a ghost town.

> I did not come here with the expectation of visiting a ghost town, this was not my motive or desire. But now I am here and you telling me about ghost tourism, I am thinking, yes, I am experiencing some of these feelings and emotions that are let's say, supernatural, paranormal. The parts of the city where there is less movement and very quiet, this is what a ghost town must be like.

Thus far, the research has identified that post-earthquake L'Aquila resembled, and from certain sections of society was labelled, a ghost town. Local residents had mixed feelings towards this term but understood why people would think it would resemble a ghost town, partly because people who had never visited before have no historical point of reference. Locals did not feel that tourists' main motivation was for ghostly experiences, but more driven by curiosity, fascination and to see what had changed over the years. As for tourists, the majority of them did not see themselves as ghost tourists, but when describing and reflecting on their experiences, many of them gave reference to ghostly feelings, being uneasy, scared, frightened. This sense of a ghostly nature was driven by various factors, such as the physical nature of the city and the ability to imagine beyond the physical environment and to use their own creative imaginations, sometimes influenced by TV and movies. Seeing that the ghostly nature is so prevalent in the city, be it overtly or inadvertently, the following explored whether locals and tourists thought that promoting the city as a ghost city would be a positive idea in attracting and managing tourism.

Opportunities for tourism: Promoting L'Aquila as a ghost town

Having established that locals could recognise and understand why terms such as ghost and dead would be attributed to the city, the researcher was curious to explore local and tourist perceptions towards the possibility of promoting ghost tourism. Following its earthquake sequence, that destroyed most of the central business district, Christchurch, New Zealand, set about offering bus tours of the affected area (Coats & Ferguson, 2013). In contrast, L'Aquila provided very little interpretation, services or facilities for tourists (Wright & Sharpley, 2018). So could the application or term ghost tourism and/or ghost tours be used to create a product in L'Aquila? Could ghost tourism be promoted as a tool for tourism? This idea was explored from the perspectives of both locals and tourists, and there was a clear contrast between them. Locals generally felt that ghost tourism was not the most suitable approach as it had a sense of entertainment to it, a lack of importance and seriousness that the city deserves. Rather, disaster tourism, earthquake related educational tourism, was deemed more suitable. The following responses capture the local perspective:

> This city has seen many tourists since the earthquake. Many people pass by and walk the streets, there is not much to see here, just the ruins and empty buildings. But people like to come. I think they are interested in seeing the damage created by the earthquake, and the reconstruction of the city, the challenges faced by the local community. I do not think ghost tourism is the more appropriate type of tourism to sell. It is not in our mentality to promote this, we have suffered too much, this is not a place for exploring ghosts, people here have suffered.

> I do not think ghost tourism offers a serious type of tourism for this city. The local people do not like people coming and taking photos of their tragedy. We need more serious forms of tourism, tourism that focuses on education and preservation, people who can bring and share knowledge of how to rebuild, people who come and then go home and tell people what really happened here in L'Aquila, and the challenges and difficulties we still have. People who will go and warn others in earthquake zones of the dangers if they are not prepared, this is more serious, more important.

Focusing on ghosts did not present a serious form of tourism for the locals. However, as identified in the research, tourists visiting L'Aquila were having ghostly experiences. The following best capture the tourists' opinions of promoting ghost tourism.

> This is a difficult question, first, I think you have to consider the local people, what do they want to promote? But having visited the city, they have not done much to cater for tourists. So they need to think of something. Having these discussions with you, I am more aware of the ghostly feelings that I have experienced here. Maybe they need to set up some tours of the city, and here they could include a ghost tour or something similar.

> Why not, there doesn't seem to be much happening here, they need to do something. This place looks like is hasn't changed much. Anything, I would like to have a tour here, someone who could guide me around. I am not sure if ghost tours are the only option, but if it is one option then yes, they could do it.

The following respondent focuses on the lack of seriousness of ghost tours, similar to the comments of the locals.

> When I think of ghost tours, I think of silly, fun places, maybe visiting a castle or a dungeon. Where people are dressed up and try to scare you. But are ghost tours too soon here. I am not sure if ghost tourism is very serious and this place looks like it is still suffering, it doesn't seem like it has moved on much since the earthquake.

Clearly, there is an issue here, the city is somewhat trapped in its past. Little has been done to establish facilities and services for tourists and locals do not feel that enough has been done to support them and their city. There also seems to be an issue of time and severity. For L'Aquila, the earthquake was devastating, and the reconstruction process has also been long and slow (Wright, 2014). Ghost tours and tourism do not necessarily come with a seriousness that the locals and tourists would want to see, but the reality is, tourists visiting this city are exposed to ghost-like experiences. The following local respondent captures a lot of this by noting the following:

> We have suffered a lot here since the earthquake. The authorities have done very little to create a serious approach to rebuilding this city, both in the buildings and the social life. We know it takes time, we all knew it would take time, but I think we were not prepared for how long. We needed more support from the start, and tourism was not a priority, local people did not want tourists visiting, we felt invaded. But now, things are different, we have had time to move on, to understand. Now, progress is still slow, but we want to grow and move on, but there is not enough support. So I think, now people might be more open to different

opportunities. I think if we can see the benefits, then maybe different tours of the city would be a good opportunity to tell our story. Maybe ghost tourism could be an opportunity.

Concluding Observations

L'Aquila became an accidental or unintentional (dark) tourism location due to the events of a devastating earthquake in 2009. It became a place where tourists could experience death and suffering. Additionally, it was established that L'Aquila could be categorised as a ghost town, and the locals, while not totally forthcoming with the term, understood why it has so often been labelled as such. Consequently, their city post-earthquake became a place where tourists could engage with ghostly and scary experiences. This was further confirmed by tourists who, when asked about their time exploring and walking in the deserted streets of the city centre, reflected on its ghostly nature. Tourists suggested that they had feelings of being frightened, spookiness, as the city had a silent and uneasy atmosphere.

Seeing that both tourists and locals understood the ghostly nature that has been associated with the city, both were asked to consider the potential of promoting the destination as a ghost town. Locals were more likely to distance themselves from such an idea, as they felt that this was not necessarily the appropriate form of tourism to attract to the city. As for the tourists, whilst they did not see themselves as ghost tourists, they felt that this could be an opportunity to attract a wider market, and consequently greater financial benefits. The issue here is that the more tourists and more movement that exists around the city, the sooner will it likely move away from the 'ghost panic' stage. However, what this research has exposed is that, due to L'Aquila's lack of a pro-active tourism approach, the city (at the time of writing) still maintains a 'ghost panic' essence, and this can be attributed to the physical emptiness of the buildings and psychological silence and limited crowded nature (movement of people) that are present. All of which are heightened when visiting at night. Potentially, for L'Aquila, there was an opportunity to utilise a section of the city in which to conduct (ghost) tours. Thus managing the tourist experience as well as ensuring locals benefitted through a more managed tourism approach that can limit the feeling of post-disaster exploitation that can come with tourists gazing on the tragedy of others (Wright, 2014).

This research has also offered further insight into the meaning of ghost towns. Importantly, it has addressed the definition of a ghost town and further established that a ghost town can show signs of life and vitality as noted by Varney (1994). L'Aquila continues to be a city in ruins, trying to recover from a tragic experience with life and movement, but large sections of the city remain a ghost town, resembling a former existence. L'Aquila is a place of death and suffering and, consequently, from an academic perspective, it can be recognised as a dark and disaster tourism location. However, this research has identified that alternative experiences, those of a ghostly nature, are present and, consequently, ghost tourism should be explored independently of other tourism concepts, as the value to tourists and locals alike requires further understanding in a place of natural catastrophe. Research also needs to explore the importance of time in establishing ghost tours and tourism experiences in disaster destinations, and the potential benefits ghost tourism could bring to a local community.

References

Agnew, P. (2016) Seven years on, shadow of earthquake still hangs over L'Aquila. *Irish Times* [online]. See https://www.irishtimes.com/news/world/europe/seven-years-on-shadow-of-earthquake-still-hangs-over-l-aquila-1.2595391.

ANSA (Agenzia Nazionale Stampa Associata) (2018) *Il centro storico dell'Aquila ancora fantasma* [online]. See http://www.ansa.it/sito/notizie/magazine/numeri/2018/04/03/il-centro-storico-dellaquila-ancora-fantasma_32ef4a91-2d00-4ecc-b6dc-fb02c8575749.html.

Babbie, E.R. (2011) *The Basics of Social Research*. Belmont, CA.: Wadsworth, Cengage Learning.

Baker, L. (2003) *Ghost Towns of Texas*. Oklahoma: Norman Publishing.

Bissell, L.J.L. (2009) *Understanding Motivation and Perception at Two Dark Tourism Attractions in Winnipeg, MB*. MA. University of Manitoba, Winnipeg.

Braun, V. and Clarke, V. (2006) Using thematic analysis in psychology. *Qualitative Research in Psychology* 3 (2), 77–101.

Brotherton, B. (2015) *Researching Hospitality and Tourism* (2nd edn). London: Sage.

Cardena, E., Lynn, S.J. and Krippner, S.C. (2000) *Varieties of Anomalous Experience: Examining the Scientific Evidence*. Washington, D.C.: American Psychological Association.

Coats, A. and Ferguson, S. (2013) Rubbernecking or rejuvenation: Post earthquake perceptions and the implications for business practice in a dark tourism context. *Journal of Research for Consumers* 23 (1), 32–65.

Daire, A. (2017) The 13 most creepy and incredible ghost towns around the world. *Independent*, 15 June [online]. See https://www.independent.co.uk/travel/the-13-most-creepy-and-incredible-ghost-towns-around-the-world-a7792041.html.

Davies, O. (2007) *The Haunted: A Social History of Ghosts*. Palgrave Macmillan: Basingstoke.

Demetri, J. (2012) Citta Morte: Ghost towns of Italy: Ancient towns and ruins in Italy. *Life in Italy*, 5 October [online]. See http://www.lifeinitaly.com/tourism/ghost-towns.asp.

Dinmore, G. (2009) Medieval town turned into ghost city. *Financial Times*, 7 April [online]. See https://www.ft.com/content/be074366-2257-11de-8380-00144feabdc0.

Dusk Till Dawn Events (2018) *Ghost Hunts* [online]. See https://www.dusktilldawnevents.co.uk/ghost-hunts-c-6.html.

DW Akademie. (2010, 6 April.) *One Year After Earthquake, L'Aquila Remains a Ghost Town*. [online]. See http://www.dw.com/en/one-year-after-earthquake-laquila-remains-a-ghost-town/a-5434698.

Eggleton, P. (2010) L'Aquila – one year on exactly. *Italy Magazine*, 6 April [online]. See http://www.italymag.co.uk/italy/laquila/laquila-one-year-exactly.

Foley, M. and Lennon, J. (1996) JFK and dark tourism: A fascination with assassination. *International Journal of Heritage Studies* 2 (4), 198–211.

Gillham, B. (2005) *Research Interviewing: The Range of Techniques*. Maidenhead: McGraw-Hill Education.

Goode, E. (1999) *Paranormal Beliefs: A Sociological Introduction*. Illinois: Waveland Press.

Hill, A. (2011) *Paranormal Media: Audiences, Spirits and Magic in Popular Culture*. New York: Routledge.

Holloway, J. (2010) Legend-tripping in spooky spaces: Ghost tourism and infrastructures of enchantment. *Environment and Planning D: Society and Space* 28 (4), 618–637

Hutt, D. (2015) Singapore's ghost hunters. *Globe Media Asia*, 9 October [online]. See http://sea-globe.com/paranormal-ghosts-singapore-southeast-asia-globe/.

Inglis, D. and Holmes, M. (2003) Highland and other haunts: Ghosts in Scottish tourism. *Annals of Tourism Research* 30 (1), 50–63.

Irwin, H.J. (2009) *The Psychology of Paranormal Belief: A Researcher's Handbook*. Hartfield: University of Hertfordshire Press.

La Repubblica. (2018) *L'Aquila, città fantasma* [online]. See http://www.repubblica.it/2006/05/gallerie/cronaca/terremoto-reportage/1.html.

Lennon, J. and Foley, M. (1999) Interpretation of the unimaginable: The U.S. Holocaust Memorial Museum, Washington, D.C., and 'dark tourism'. *Journal of Travel Research* 38 (1), 46–50.

Long, J. (2008) *Researching Leisure, Sport and Tourism. The Essential Guide*. London: Sage.

Mason, P. (2014) *Researching Tourism, Leisure and Hospitality for your Dissertation*. Oxford: Goodfellow.

McLaughlin, T. (2010) L'Aquila after the earthquake. *Italy Magazine*, 20 January [online]. See http://www.italymag.co.uk/italy-featured/laquila-province/l-aquila-after-earthquake.

Miklós, V. (2013) The strangest and most tragic ghost towns from around the world. *Gizmodo Media Group*, 13 September [online]. See https://io9.gizmodo.com/the-strangest-and-most-tragic-ghost-towns-from-around-t-1308304680.

Molle, A. and Bader, C. (2014) Paranormal science from America to Italy: A case of cultural homogenisation. In O. Jenzen and S.R. Munt (eds) *The Ashgate Research Companion to Paranormal Cultures* (pp. 121–138). London: Ashgate.

Puliafito, A. (2011) *Intorno a L'Aquila paesi fantasma e zone rosse*. [Blog] IlFattoQuotidiano.it. See https://www.ilfattoquotidiano.it/2011/04/24/intorno-a-laquila-paesi-fantasma-e-zone-rosse/106593/.

Rittichainuwat, B. (2011) GHOSTS: A travel barrier to tourism recovery. *Annals of Tourism Research* 38 (2), 437–459.

Santilli, de-B. (2013) L'Aquila: città fantasma che vuole rinascere. *ANSA*, 5 April. See http://www.ansa.it/web/notizie/rubriche/speciali/2013/04/03/Aquila-quattro-anni-dopo-citta-fantasma_8506711.html.

Sharp, R. (2010) New and improved: Rebuilding a disaster zone. *Independent*, 9 April [online]. See https://www.independent.co.uk/arts-entertainment/architecture/new-and-improved-rebuilding-a-disaster-zone-1948094.html.

Sharpley, R. (2005) Travels to the edge of darkness: Towards a typology of 'dark tourism'. In C. Ryan, S.J. Page and M. Aicken (eds) *Taking Tourism to the Limits: Issues, Concepts and Managerial Perspectives* (pp. 215–226). London: Elsevier.

Sharpley, R. and Stone, P.R. (2009) *The Darker Side of Travel: The Theory and Practice of Dark Tourism*. Bristol: Channel View Publications.

Spinelli, B. (2012) L'Aquila, la città che non c'è più invasa da New Town fantasma. *La Repubblica*, 6 June [online]. See http://www.repubblica.it/cronaca/2012/06/06/news/l_aquila_macerie-36626978/.

Stone, P.R. (2006) A dark tourism spectrum: Towards a typology of death and macabre related sites, attractions and exhibitions. *Tourism: An Interdisciplinary International Journal* 54 (2), 145–160.

Taylor, L. (2011) The reality of ghosts: Explaining our fascination with the supernatural. *The Open University*, 17 May [online]. See http://www.open.edu/openlearn/history-the-arts/culture/the-reality-ghosts-explaining-our-fascination-the-supernatural.

The Telegraph (2018, 10 October) *The World's Most Haunted Hotels* [online]. See https://www.telegraph.co.uk/travel/hotels/galleries/The-worlds-most-haunted-hotels/.

Tripadvisor (2018) *10 Top Ghost Tours Around the World* [online]. See https://www.tripadvisor.co.uk/TripNews-a_ctr.ghosttoursEN.

UNFrame (2015, 13 April) *L'Aquila, Ghost City for Six Years and On. Italy* [online]. See http://unframe.com/laquila-ghost-city-for-six-years-and-on-italy/.

Varney, P. (1994) *Southern California's Best Ghost Towns: A Practical Guide*. Oklahoma: Norman Publishing.

Virgin Experience Days (2018) *Ghost Hunting* [online]. See https://www.virginexperiencedays.co.uk/ghost-hunting.

Watt, C. and Wiseman, R. (2009) Foreword. In H.J. Irwin (ed.) *The Psychology of Paranormal Belief* (pp. vii–viii). Hartfield: University of Hertfordshire Press.

Weiser, K. (2010) Ghost towns. Types & code of ethics. *Legends of America* [online]. See https://www.legendsofamerica.com/gt-ghosttownethics/

Wright, D.M.W. (2014) Residents' Perceptions of Dark Tourism Development: The Case of L'Aquila, Italy. Unpublished PhD thesis, University of Central Lancashire.

Wright, D.W.M. (2016) Hunting humans: A future for tourism in 2200. *Futures* 78–79, 34–46.

Wright, D. and Sharpley, R. (2018) Local community perceptions of disaster tourism: The case of L'Aquila, Italy. *Current Issues in Tourism* 21 (14), 1569–1585.

11 Conclusion: Earthquakes and Tourism – An Emerging Research Agenda

Girish Prayag and C. Michael Hall

This book has highlighted the complex nature of earthquakes and the impacts they may have on individuals, communities and countries. Although largely negative, these impacts can be long lasting and have to be managed for the successful recovery of destinations, communities and organizations. Through its various chapters, this book provides a global coverage of some of the most devastating earthquakes we have seen in the last few decades. From the devastating Sumatran earthquake and tsunami of 2004 (Subadra, this volume), the 2010 Haiti earthquake (Morpeth, this volume), the 2015 earthquake in Nepal (Das & Chakrabarty, this volume), the 2009 earthquake in L'Acquila (Wright, this volume) and to the 2010–2011 Canterbury earthquakes (Amore, this volume), it is clear that both developed and developing countries are heavily impacted by such disasters. What emerges as a common problem facing the various destinations and communities investigated in this book is the lack of coordination between various agencies post-quake and often the lack of disaster preparedness, despite these communities being in high risk areas and hence vulnerable to negative impacts. Issues of resilience and sustainability also emerge from the various chapters as being key concerns for destinations and communities post-quake (Hall *et al.*, 2018). This concluding chapter examines some of these issues in relation to the extant literature on disasters and tourism.

As discussed in Chapter 1 (Hall & Prayag, this volume), earthquakes are the only disasters that can bring an onset of other deadly disasters such as tsunamis and landslides. Irrespective of the stage of economic development, earthquakes have profound effects on the economic and social fabric of countries and communities. While many existing studies (e.g. Huan *et al.*, 2004; Huang *et al.*, 2008; Wu

& Hayashi, 2014; Yang et al., 2011) examine negative impacts, these also provide opportunities for the impacted communities to review and assess their disaster preparedness initiatives, if any. Indeed, a major issue in the analysis of earthquake disasters and tourism is the lack of long-term studies that cover all stages of the disaster recovery process as well as the extent to which perceptions of earthquake risk together with disaster management policies and earthquake preparedness go through issue-attention cycles (Hall, 2002, 2010). In addition, there is a lack of long-term studies of the tourist who experience disasters such as earthquakes and the implications it has for their longer term perception of a destination , willingness to revisit and word of mouth.

Disaster recovery is a long-term process that requires the input and effort of multiple stakeholders and therefore provides significant opportunities for destinations and communities to reassess existing relationships, processes and governance structures in place to rebuild something better. For the tourism industry in particular, it is an opportunity to revisit sustainability and resilience initiatives in place or to set up new ones. It is an opportunity to develop new market segments such as eco-tourism (Wang & Cater, 2015), revamp and rebrand destination image (Ketter, 2016) and rebuild hotels and convention centres and other attractions and infrastructure. Although such developments are not necessarily without controversy (Amore & Hall, 2016a; Hall & Amore, 2019), Amujo and Otubanjo (2012) showed that the post-disaster context can help with nation rebranding. It is also an opportunity to attract investments and new migrants to rebuild the tourism industry. As highlighted in Chapter 2 (Martini & Platania, this volume), tourism organizations often represent a core component of the resilience of destination communities. Through their ability to create new products relatively quickly and target new customers, tourism organizations have a significant role to play in the community 'bouncing back' post-quake. Yet, only limited studies have examined what new products, attractions and segments emerge from post-disaster tourism (Biran et al., 2014; Mair et al., 2016).

Chapter 2 (Martini & Platania, this volume) also highlights the complex relationship between the vulnerability of destinations and the resilience of tourism enterprises. The authors argue that disaster preparedness should look beyond the vulnerability of infrastructure and include strategies that reduce social vulnerability. Recent studies on tourism and destination resilience (Amore et al., 2018; Calgaro et al., 2014; Cochrane, 2010) have argued that the socioecological approach to studying the resilience of the tourism system explicitly accounts for the role of people and communities and their contribution to the resilience of the system. In this way, resilience building is also about reducing social vulnerability due to unexpected changes in the environment. As noted in Chapter 1 (Hall & Prayag, this volume), the wellbeing of individuals often gets sidelined in disaster recovery plans at the expense of infrastructure rebuild. This has led to the call by a number of researchers for a more holistic approach in managing the recovery phase (Ritchie, 2004; Mannakkara & Wilkinson, 2014; Hall et al., 2016). Without a long term view of the well-being of residents being included in any disaster preparedness and recovery plans, there will be missed opportunities to strengthen social capital. As suggested in many previous studies, the recovery phase is an opportunity to strengthen and rebuild social capital (Aldrich & Meyer, 2015; Nakagawa & Shaw, 2004). Importantly, social capital not only contributes to the resilience of communities but also the resilience of tourism organizations (Chowdhury et al., 2019). Nevertheless, such a situation also creates a substantial challenge to governments and the tourism industry at destinations with respect to the contribution they make to communities with respect to not only employment but also wage levels and the extent to which tourism gives back to a community rather than overwhelm it.

Collectively, the various chapters also highlight the lack of preparedness of tourism businesses in the face of imminent disaster threats. Often tourism businesses understand the significance of the threat and/or the extent to which an earthquake can impact their business but as shown by the results of the survey conducted among tourism businesses near Mount Etna, Italy, they are not necessarily proactive in managing the potential risks (see Chapter 2, Martini & Platania, this volume). This emerges as a significant issue in the literature (see Hystad & Keller, 2006; Orchiston, 2013; Mair et al., 2016), with the conclusion that tourism businesses can be impacted severely by disasters due to their reactive approaches to disaster preparedness. They may have plans in place but not implement or update those plans to reflect the changing nature of the business environment.

Nevertheless, despite the well-accepted notion that preparedness, response and recovery are related, this process is neither sequential nor linear. Most of the chapters in the book tend to argue the importance of disaster preparedness for tourism businesses and communities that depend on tourism. Yet there seems to be no unified approach that can be used to make businesses and communities more prepared for and resilient to earthquakes and other disasters. Certainly Chapter 3 (Das & Chakrabarty, this volume) hones in on the importance of communication during and after an earthquake, with the authors pointing to the need for various stakeholders to coordinate communication activities post-quake and the role of social sensing and social sensing systems in managing the response and recovery phases of disasters. This is especially significant with respect to the health and medical needs of tourists, residents and industry employees following an earthquake. From a public health perspective addressing injuries and the potential spread of disease is clearly important. Destination organizations and tourism businesses are also seeking to evacuate tourists as soon as possible from the affected area. This is partly because of concerns over wellbeing but it is also so as to minimize costs and reduce the possibilities of negative publicity because of coverage of tourist complaints on social and general media. Mair *et al.* (2016) also pinpoint to the tourism industry needing a more coordinated approach for communicating with tourists during and after a disaster (see also Subadra, this volume). Having a disaster marketing and communication plan in place may help to assign clear responsibilities and clarify the role of various stakeholders in the response and recovery phases (Hystad & Keller, 2008). However, there is a need for greater evaluation of the appropriateness of these plans in the research literature especially after an earthquake has occurred.

In particular, previous studies have highlighted that the media can be a double-edged sword post-disaster (Robinson & Jarvie, 2008; Yang *et al.*, 2011). While an important communication channel for organisations and destinations, the media can also facilitate negative, long-lasting impressions amongst tourists (Huan *et al.*, 2004; Huang & Min, 2002; Yang *et al.*, 2011). Familiarization visits and events can be used to counteract negative media coverage after disasters (Huang & Min, 2002). Robinson and Jarvie (2008) note that communication efforts need to not only articulate the readiness of the post-disaster destination's infrastructure but also the citizens' psychological capacity to accept visitors. As shown in the work of Orchiston and Higham (2016), the destination has to go through a demarketing phase (Hall, 2014) to allow the basic infrastructure to be rebuilt which can then be followed by campaigns emphasizing how the destination has 'bounced back' from the disaster. Communication campaigns need to reassure tourists who may be concerned that it is 'too soon' to visit due to the psychological devastation caused by the disaster and that it may not be safe to visit. Nevertheless, there is often insufficient attention given to the VFR market segment as both a market willing to return to a destination early because of its commitment to people and place, and which will not necessarily put undue stress on affected infrastructure, while also generating economic and social capital in the affected destination. Research also indicates that different communication campaigns may be required for natural disasters compared to acts of terrorism, the latter being more difficult to counter with positive media messages (Robinson & Jarvie, 2008).

A significant component of many disaster management models is mitigation strategies (Huan *et al.*, 2004; Paton *et al.*, 2000; Ritchie, 2008). Mitigation strategies are critical to lessen the impact of disasters on businesses and communities. As suggested in Chapter 4 (Subadra, this volume), despite a destination having disaster mitigation strategies in place, stakeholders may not always have the competencies to implement these strategies. The lack of coordination from ground level to government can hinder the implementation of mitigation plans and/or their effectiveness. This is especially the case if tourism industry disaster response is separate from that of disaster management agencies. The literature suggests five broad areas around which mitigation strategies can be crafted, including environmental management, land-use planning, protection of critical facilities, networks and partnerships, and financial capital (Ritchie, 2008). Prayag and Orchiston (2016) provide an overview of 17 specific mitigation strategies that were used by businesses following the Canterbury earthquakes. Similar to Ritchie's (2008) work, these strategies included managing available cash and credit, insurance claims, having well designed and well-built buildings, the use of slack resources (e.g. people, equipment) and, more importantly, practicing response to a disaster. Prayag and Orchiston (2016) also highlight the importance of building relationships with staff, banks and lenders and more broadly, relationships with neighbours. In essence, as argued earlier, social capital as a mitigation strategy has its merits.

Chapters 2 (Martini & Platania, this volume) and 5 (Amore, this volume) also discuss issues of trust and governance between tourism businesses and regional government, highlighting how institutional frameworks can hinder rather than assist tourism businesses with disaster preparedness and recovery strategies. Several studies argue that recovery is stalled by poor governance structures (Amore & Hall, 2016b; Dimmer, 2014; Hall *et al.*, 2016). Collective and sustainable recovery and disaster risk reduction, which is deemed socially and morally acceptable, is the result of negotiation and contestation between different stakeholders (Larsen *et al.*, 2011). These stakeholders may have very different views, for example, of resilience and sustainability (Hall *et al.*, 2018). Research on the aftermath of Hurricane Katrina (Gotham, 2007, 2012; Gotham & Greenberg, 2008, 2014) and the Christchurch earthquake sequence (Hayward, 2013; Amore & Hall, 2016b; Amore *et al.*, 2017; Hall & Amore, 2019), for example, strongly illustrates the way in which different political and economic interests view rebuild and recovery in terms of both decision-making processes and actual outcomes. Amore continues his analysis of the Christchurch situation in this volume (Chapter 5) and suggests that the approach used for recovery is often *ad hoc* and piecemeal, with no single entity having a holistic view of the process. Consequently, decisions may be sub-optimal and not well aligned with expectations of the community with decisions instead driven by the interests of central government and business, especially in terms of real estate values and particular ways of framing a 'competitive' city. Chapter 5 (Amore, this volume) raises the question of what type of meta-governance is required for successful recovery of tourist destinations (Amore & Hall, 2016a, 2017). The alignment of the values and interests of recovery authorities with those of tourism-relevant stakeholders seem to be what the author suggests as being necessary for policy making and advisory powers to be of added value in the recovery phase.

Throughout the book, many chapters recognize that tourism businesses are often not prepared for disasters and therefore have limited ideas on how to bounce back. Previous studies have found that small tourism businesses often do not have the simplest disaster response tool such as an emergency kit (Orchiston, 2013). Chapter 6 (Morpeth, this volume) highlights how the lack of preparedness, poverty and undemocratic political systems have a huge impact on the human cost of disasters. Yet having a plan in place does not guarantee faster and better recovery. Corey and Deitch (2011) found that having plans in place before Hurricane Katrina was not correlated with business recovery after the disaster. One major issue is that communities, agencies and organizations do not always learn from the experience of previous disasters, as argued in Chapter 6 (Morpeth, this volume). Studies have also evaluated the effects of prior disaster experience on government-level preparedness. Alexander (2013) investigated the extent to which the Italian government learned from the 1980 Irpinia earthquake and adjusted accordingly when responding to the 2009 L'Aquila earthquake. While the 2009 earthquake was more moderate than the 1980 quake, its effects were worsened by the failure of the government to implement recommended changes to the building code and to encourage business continuity plans. Hence, it seems that knowledge management and transfer in terms of capturing and dissemination of previous learning from disasters remain a problematic area that certainly warrants more attention from researchers (Orchiston & Higham, 2016; Mair et al., 2016), and which also perhaps reflects concerns over metagovernance in earthquake affected destinations (Amore, Chapter 5, this volume). Related to issues of disaster preparedness and knowledge management, Chapter 7 (Hashimoto & Telfer, this volume) in particular, examines the role of interagency collaboration for tourism crisis management and the importance of organizational learning in the recovery phase.

Chapter 8 (Mardiah et al., this volume) reinforces this idea of interagency collaboration and highlight the tensions and difficulties in setting policies that will stimulate business recovery when the priority is often housing rehabilitation and reconstruction. The chapter highlights the challenges faced by small and medium tourism enterprises involved in handicraft manufacturing for tourists. This theme of SMEs is discussed further in Chapter 9 (Fang et al., this volume) where the authors examine the resilience of owner-managers of SMEs post-quake. The chapter argues that there is a close relationship between the psychological resilience and organizational resilience in the context of SMEs. Entrepreneurial resilience contributes positively toward the tourism industry bouncing back from a disaster. Hence, the authors make the argument for disaster management models to incorporate long-term well-being of individuals and communities in a more explicit way. The interaction of different types and level of resilience has been identified as a significant gap in the tourism literature (Hall *et al.*, 2016, 2018; Prayag *et al.*, 2019).

While most of the chapters in this book examine supply side issues arising from earthquakes, Chapter 6 (Morpeth, this volume) questions the role of tourists post-quake. While several studies have examined how post-quake destinations become famous for their dark tourism or thanatourism attractions, the ethical issues that such sites raise among residents and survivors remain overlooked. Chapter 6 (Morpeth, this volume) questions the gaze of cruise tourists following the Haiti earthquake. Chapter 10 (Wright, this volume) builds on the idea of thanatourism and examines the emergence of ghost towns as tourist attractions following the earthquake in L'Acquila. The experience of tourists (international or domestic) or the demand side of dark tourism as well as the supply side is well researched (Light, 2017) but often the views of local community, including residents have been ignored (Kim & Butler, 2015; Lin *et al.*, 2018; Prayag, 2016). Dark tourism remains one particular form of tourism that can emerge after a disaster (Biran *et al.*, 2014). Tourism recovery for destinations affected by disasters can be facilitated by dark tourism, given the newly formed dark attributes that emerge from the disaster (Biran *et al.*, 2014). This new segment can potentially take a significant role in disaster recovery management and make a contribution to all stages of disaster recovery (Muskat *et al.*, 2015). However, its use is not without opposition given concerns as to the image it may provide of a destination. Nevertheless, it may be that, depending on the extent of recovery, dark tourism may only have a limited lifespan in post-earthquake contexts apart from those locations that are never rebuilt.

In seeking to build a research agenda for earthquakes and tourism several themes appear to be central to future research efforts. First, the need for sustained longitudinal research. Second, a closer examination of the effects that earthquakes have on the interaction between the various elements of the tourism system as well as the wider environment within which tourism occurs. Third, the relationships between the tourism industry and disaster management organizations with respect to preparedness and response. This also includes the role that tourism marketing can play. Fourth, the role of government and political and economic interests in shaping the trajectory of destination recovery including both the tangible, i.e. infrastructure and facilities, and intangible, i.e. branding and image, elements of destinations. Fifth, and finally, the development of improved models and frameworks with which to better understand earthquake related resilience at different scales. We hope that this book has contributed in some small way to advancing these areas and improving the welfare of those affected by earthquakes.

References

Aldrich, D.P. and Meyer, M.A. (2015) Social capital and community resilience. *American Behavioral Scientist* 59 (2), 254–269.

Alexander, D. (2013) An evaluation of medium-term recovery processes after the 6 April 2009 earthquake in L'Aquila, Central Italy. *Environmental Hazards* 12 (1), 60–73.

Amore, A. and Hall, C.M. (2016a) From governance to meta-governance in tourism? Re-incorporating politics, interests and values in the analysis of tourism governance. *Tourism Recreation Research* 41 (2), 109–122.

Amore, A. and Hall, C.M. (2016b) 'Regeneration is the focus now': Anchor projects and delivering a new CBD for Christchurch. In C.M. Hall, S. Malinen, R. Vosslamber and R. Wordsworth (eds) *Business and Post-disaster Management: Business, Organisational and Consumer Resilience and the Christchurch Earthquakes* (pp. 181–199). Abingdon: Routledge.

Amore, A. and Hall, C.M. (2017) National and urban public policy agenda in tourism. Towards the emergence of a hyperneoliberal script? *International Journal of Tourism Policy* 7 (1), 4–22.

Amore, A., Hall, C.M. and Jenkins, J.M. (2017) They never said 'Come here and let's talk about it': Exclusion and non-decision making in the rebuild of Christchurch, New Zealand. *Local Economy* 32 (7), 617–639.

Amore, A., Prayag, G. and Hall, C.M. (2018) Conceptualizing destination resilience from a multilevel perspective. *Tourism Review International* 22 (3–4), 235–250.

Amujo, O.C. and Otubanjo, O. (2012) Leveraging rebranding of 'unattractive' nation brands to stimulate post-disaster tourism. *Tourist Studies* 12 (1), 87–105.

Biran, A., Liu, W., Li, G. and Eichhorn, V. (2014) Consuming post-disaster destinations: The case of Sichuan, China. *Annals of Tourism Research* 47, 1–17.

Calgaro, E., Lloyd, K. and Dominey-Howes, D. (2014) From vulnerability to transformation: A framework for assessing the vulnerability and resilience of tourism destinations. *Journal of Sustainable Tourism* 22 (3), 341–360.

Chowdhury, M., Prayag, G., Orchiston, C. and Spector, S. (2019) Post-disaster social capital, adaptive resilience and business performance of tourism organizations in Christchurch, New Zealand. *Journal of Travel Research* 58 (7), 1209–1226.

Cochrane, J. (2010) The sphere of tourism resilience. *Tourism Recreation Research* 35 (2), 173–185.

Corey, C.M. and Deitch, E.A. (2011) Factors affecting business recovery immediately after Hurricane Katrina. *Journal of Contingencies and Crisis Management* 19 (3), 169–181.

Dimmer, C. (2014) Evolving place governance innovations and pluralising reconstruction practices in post-disaster Japan. *Planning Theory & Practice* 15 (2), 260–265.

Gotham, K.F. (2007) (Re) Branding the Big Easy: Tourism rebuilding in Post-Katrina New Orleans. *Urban Affairs Review* 42, 823–850.

Gotham, K.F. (2012) Disaster, Inc.: Privatization and post-Katrina rebuilding in New Orleans. *Perspectives on Politics* 10, 633–646.

Gotham, K.F. and Greenberg, M. (2008) From 9/11 to 8/29: Post-disaster recovery and rebuilding in New York and New Orleans. *Social Forces* 87, 1039–1062.

Gotham, K.F. and Greenberg, M. (2014) *Crisis Cities: Disaster and Redevelopment in New York and New Orleans*. Oxford: Oxford University Press.

Hall, C.M. (2002) Travel safety, terrorism and the media: The significance of the issue-attention cycle. *Current Issues in Tourism* 5 (5), 458–466.

Hall, C.M. (2010) Crisis events in tourism: Subjects of crisis in tourism. *Current Issues in Tourism* 13 (5), 401–417.

Hall, C.M. (2014) *Tourism and Social Marketing*. Abingdon: Routledge.

Hall, C.M. and Amore, A. (2019) The 2015 Cricket World Cup in Christchurch. *Journal of Place Management and Development*. See https://doi.org/10.1108/JPMD-04-2019-0029.

Hall, C.M., Prayag, G. and Amore, A. (2018) *Tourism and Resilience: Individual, Organisational and Destination Perspectives*. Bristol: Channel View Publications.

Hall, C.M., Malinen, S., Vosslamber, R. and Wordsworth, R. (eds) (2016) *Business and Post-disaster Management: Business, Organisational and Consumer Resilience and the Christchurch Earthquakes*. Abingdon: Routledge.

Hayward, B.M. (2013) 'Rethinking resilience: Reflections on the earthquakes in Christchurch, New Zealand, 2010 and 2011. *Ecology and Society* 18, 37–42.

Huan, T.C., Beaman, J. and Shelby, L. (2004) No-escape natural disaster: Mitigating impacts on tourism. *Annals of Tourism Research* 31 (2), 255–273.

Huang, J.H. and Min, J.C.H. (2002) Earthquake devastation and recovery in tourism: The Taiwan case. *Tourism Management* 23 (2), 145–154.

Huang, Y.C., Tseng, Y.P. and Petrick, J.F. (2008) Crisis management planning to restore tourism after disasters: A case study from Taiwan. *Journal of Travel & Tourism Marketing* 23 (2–4), 203–221.

Hystad, P. and Keller, P. (2006) Disaster management: Kelowna tourism industry's preparedness, impact and response to a 2003 major forest fire. *Journal of Hospitality and Tourism Management* 13 (1), 44–58.

Hystad, P.W. and Keller, P.C. (2008) Towards a destination tourism disaster management framework: Long-term lessons from a forest fire disaster. *Tourism Management* 29 (1), 151–162.

Larsen, R.K., Calgaro, E. and Thomalla, F. (2011) Governing resilience building in Thailand's tourism-dependent coastal communities: Conceptualising stakeholder agency in social–ecological systems. *Global Environmental Change* 21 (2), 481–491.

Light, D. (2017) Progress in dark tourism and thanatourism research: An uneasy relationship with heritage tourism. *Tourism Management* 61, 275–301.

Ketter, E. (2016) Destination image restoration on Facebook: The case study of Nepal's Gurkha Earthquake. *Journal of Hospitality and Tourism Management* 28, 66–72.

Kim, S. and Butler, G. (2015) Local community perspectives towards dark tourism development: The case of Snowtown, South Australia. *Journal of Tourism and Cultural Change* 13 (1), 78–89.

Lin, Y., Kelemen, M. and Tresidder, R. (2018) Post-disaster tourism: Building resilience through community-led approaches in the aftermath of the 2011 disasters in Japan. *Journal of Sustainable Tourism* 26 (10), 1766–1783.

Mair, J., Ritchie, B.W. and Walters, G. (2016) Towards a research agenda for post-disaster and post-crisis recovery strategies for tourist destinations: A narrative review. *Current Issues in Tourism* 19 (1), 1–26.

Mannakkara, S. and Wilkinson, S. (2014) Re-conceptualising 'Building Back Better' to improve post-disaster recovery. *International Journal of Managing Projects in Business* 7 (3), 327–341.

Muskat, B., Nakanishi, H. and Blackman, D. (2015) Integrating tourism into disaster recovery management. In K. Campiranon and B.W. Ritchie (eds) *Crisis and Disaster Management in the Asia-Pacific* (pp. 97–115). Wallingford: CABI Publishing.

Nakagawa, Y. and Shaw, R. (2004) Social capital: A missing link to disaster recovery. *International Journal of Mass Emergencies and Disasters* 22 (1), 5–34.

Orchiston, C. (2013) Tourism business preparedness, resilience and disaster planning in a region of high seismic risk: The case of the Southern Alps, New Zealand. *Current Issues in Tourism* 16 (5), 477–494.

Orchiston, C. and Higham, J.E.S. (2016) Knowledge management and tourism recovery (de) marketing: The Christchurch earthquakes 2010–2011. *Current Issues in Tourism* 19 (1), 64–84.

Paton, D., Smith, L. and Violanti, J. (2000) Disaster response: Risk, vulnerability and resilience. *Disaster Prevention and Management: An International Journal* 9 (3), 173–180.

Prayag, G. (2016) It's not all dark! Christchurch residents' emotions and coping strategies with dark tourism sites. In C.M. Hall, S. Malinen, R. Vosslamber and R. Wordsworth (eds) *Business and Post-disaster Management: Business, Organisational and Consumer Resilience and the Christchurch Earthquakes* (pp. 155–166). Abingdon: Routledge.

Prayag, G. and Orchiston, C. (2016) Earthquake impacts, mitigation, and organisational resilience of business sectors in Canterbury. In C. M. Hall, S. Malinen, R. Vosslamber and R. Wordsworth (eds) *Business and Post-disaster Management: Business, Organisational and Consumer Resilience and the Christchurch Earthquakes* (pp. 111–132). Abingdon: Routledge.

Prayag, G., Spector, S., Orchiston, C. and Chowdhury, M. (2019) Psychological resilience, organizational resilience and life satisfaction in tourism firms: Insights from the Canterbury earthquakes. *Current Issues in Tourism*. See https://doi.org/10.1080/13683500.2019.1607832.

Ritchie, B.W. (2004) Chaos, crises and disasters: A strategic approach to crisis management in the tourism industry. *Tourism Management* 25 (6), 669–683.

Ritchie, B. (2008) Tourism disaster planning and management: From response and recovery to reduction and readiness. *Current Issues in Tourism* 11 (4), 315–348.

Robinson, L. and Jarvie, J.K. (2008) Post-disaster community tourism recovery: The tsunami and Arugam Bay, Sri Lanka. *Disasters* 32 (4), 631–645.

Wang, C.C. and Cater, C (2015) Ecotourism as a sustainable recovery tool after an earthquake. In B.W. Ritchie and K. Campiranon (eds) *Tourism Crisis and Disaster Management in the Asia-Pacific* (pp. 209–226). Wallingford: CABI.

Wu, L. and Hayashi, H. (2014) The impact of disasters on Japan's inbound tourism demand. *Journal of Disaster Research* 9 (sp), 699–708.

Yang, W., Wang, D. and Chen, G. (2011) Reconstruction strategies after the Wenchuan Earthquake in Sichuan, China. *Tourism Management* 32 (4), 949–956.

Index

accessibility 10, 122
adaptation 7, 8, 21
    individual 8
    psychological 7
accommodation capacity 12, 91
adaptive capacity 19–20, 23, 40, 159, 161, 162
adaptive cycle 20
adaptive systems 21, 23
Africa 113
Asia 101, 102
Asian Development Bank 143
Association of Southeast Asian Nations (ASEAN) 52, 54, 58
attractions 78, 122–3
Australia 76, 91, 93, 156, 146, 171, 173
    Sydney 171

behavioural interventions 164; *see also* social marketing
Boxing Day tsunami *see* Indian Ocean Boxing Day tsunami
branding 66, 84, 125, 149, 189, 199
    re-branding 84, 91, 92
bus 106, 123, 129
    tours 172, 173, 187
bushfire *see* wildfire
business 157–8, 162–4
    adaptive capacity 159
    difficulties post-earthquake 145
    recovery 23, 85, 144–6, 148, 159, 164, 198
    resilience 46
business continuity management and planning 143, 148, 150, 198
business environment 46, 195
business stakeholders 90

Canada 144, 173
    Winnipeg 173
capacity 5, 12, 16, 22, 36, 100, 149, 150
    adaptive 19–20, 23, 40, 154, 149, 163
    building 147
    community 19, 140
    economic 4
    organizational 24

psychological 154, 160–1, 196
    and resilience 154
    response 37, 71
    transformative 40
car 129
Caribbean 101, 106
Carnival Cruises 107
change 20–2, 36–8
    dynamics 19
    and resilience 24–5
Chile 2
    Concepcion 2
    Metropolitana 2
    Rancagua 2
    Talca 2
    Temuco 2
    Valparaiso 2
China 3, 14, 58, 133
    Sichuan 14
Christchurch Cathedral 88
coastal areas 65, 70, 70, 77
climate change 100–1
collaboration 9, 12, 20, 61, 91–2, 94, 116–17, 133, 159, 198
    and crisis knowledge management 117
community belonging 46
community preparedness 45
complex adaptive system 21
complexity 19, 27, 93, 104
connectivity 21, 56
cruise tourism 105–7, 112
Cyprus 85

dark tourism 14–15, 18, 172–3, 177, 189, 199
destination 10, 140
    appeal 88
    image 17, 51, 52, 134, 194
    planning 20
    policies 48
    recovery 13, 17, 117, 173, 199
    resilience 20–1, 25, 194
destination image restoration theory 130
destination management 53–4, 56–7

Index

destination marketing 10
destination vulnerability 20
disaster capitalism 87
disaster
    defined 102
    recovery 5
    resilience 12, 20
    response 10, 12, 15, 19, 27, 51, 60, 116–17, 197, 198
    risk reduction 20, 59, 197
disaster management cycle 26
disaster tourism 102–3
    ethics 103–4
disaster relief 10, 22, 99, 104, 116, 122, 126–7, 133
Dominican Republic 101–2, 107
drought 41, 42, 100

earthquake 1–5, 25, 43–7, 87, 120, 144
    defined 3–5
    global human impact 1–3
    impacts on tourism system 11
    origins 3–5
    preparedness 43, 47
    recovery 7, 87, 120, 144
    swarms 5
earthquake risk 3, 18, 43–4, 194
    information sources 44
economic development 106, 139, 143, 149, 155, 193
employment 106, 107, 149, 195
enterprise resilience 154
environmental change 36
Europe 39, 113, 173
European airspace 112
European Union 143
Eyjafallajokull volcano eruption 112

Facebook 55, 56, 59, 60, 126, 127, 130–1; *see also* social media
flood 41, 58, 174

gender 106
geopolitics 106
Germany 146
ghost tourism 168–90
    cities 174
ghost town 175–6, 180, 189
global warming *see* climate change
globalization 78
governance 12, 25, 39, 84–6, 93, 150, 194, 197
    disaster risk 112, 144
    tourist areas 84, 94–5
    urban redevelopment 83, 93

Haiti 3, 26, 99–114
    Labadee 105, 106–7, 108, 109
    Port-au-Prince 107, 109
health 5, 7–8, 24, 59, 101, 195
    mental 8
    services 56
health tourism 122–3, 125–6
heritage 4, 15, 18, 39, 58, 88, 141; *see also* World Heritage
Honduras 84
hospitality 82–3, 85, 88, 89, 90, 93, 127, 133
hot springs 125–6
human resources 57, 60
Hurricane Katrina 19, 22, 36, 84, 114, 197, 198

ice dome 126
Iceland 112–13
India 4
    Ross Island 174
Indian Ocean 4
Indian Ocean Boxing Day tsunami 4, 15, 22, 27
Indonesia 3, 4, 26, 65–81, 138–52
    Bali 65–81, 142
    Jakarta 76, 138
    Java 66, 138
    Jogjakarta 138–52
    Lombok 66, 70, 142
    Sumatra 4
informal sector 156
innovation 21, 87, 129, 140, 153, 159, 164
Instagram 55; *see also* social media
insurance 2
interest groups 83
Ionian Sea 39
Iran 3
Israel 164
Italy 26, 39–49, 168, 175–90, 195
    Catania 39
    Craco 174
    L'Aquila 168, 175–90
    Sicily 39
    Umbria 14

Japan 2, 115–34
    Aomori 2
    Aso 2
    Chiba 2
    Chuo Ward 2
    Fukuoka 2
    Fukushima 2, 5
    Hashima Island 174
    Hyogo 2

Kobe 2
Kyoto 2
Kumamoto 2, 115–34
Lbaraki 2
Lwate 2
Mashiki 2
Minamiaso 2
Miyagi 2
Miyazaki 2
Oita 2
Osaka 2
Tochigi 2
Tokyo 2
Yamagata 2
Yamaguchi 2

Korea *see* South Korea

landslide 3, 4, 5, 58, 76, 115, 121, 126, 129–31, 132, 174, 193
lava 39, 121
Less Developed Countries (LDCs) 52, 56–7, 61
lateral spreading 86, 120
liquefaction 5, 86
Lisbon earthquakes and tsunami 4

marketing 10, 15, 92, 142, 145–6, 173, 196; *see also* branding, promotion, social media
    communication 15, 173, 196
    destination 10
    strategies 17, 22, 196
Mexico 2
    Mexico City 2
    Morelos 2
    Puebla 2
Mount Agung 69, 78
Mount Etna 25, 39–49, 195
Mount Everest 57
Mount Saint Helens 102

Namibia 174
    Kolmanskop 174
national park 39, 122–3
natural disaster *see* disaster
Nepal 14, 17, 19, 27, 51–2, 57–64
    Bhaktapur 58
    Kathmandu 58
    Patan 58
Nepal Tourism Board 59
Netherlands 143, 146
New Zealand 2, 5, 8, 14, 17, 26, 40, 82–95, 153–64, 173, 187
    Canterbury 2, 8, 17, 26
    Christchurch 2, 17, 26, 82–95, 153, 187
    Kaikōura 2, 26, 153–64
    Lyttleton 2
    Marlborough 2
    Picton 2, 153
    Timaru 2
    Wellington 2
non-place 105–7
normality 21, 113
North Macedonia 3

onsen baths 125–6
Oxfam 110

promotion 14, 17, 66, 91, 123, 146, 149
pro-poor tourism 112
protected area *see* national park
public policy 83
pyroclastic flow 121

rail 115, 121–2, 127, 130, 153, 173
reef tourism 156
resilience 9, 20–5, 36–40, 154–63, 194, 198
    community-based 6, 9, 20, 21–2, 36, 39, 99
    destination 20, 21, 25, 26, 194
    definition 157
    employee 23, 154, 159
    entrepreneurial 156–63, 198
    individual 157
    organisational 154–63, 198
    personal 163
    psychological 157, 161–3
    social 38, 40
    societal 37
    socio-ecological 36
    and sustainability 24–5
risk 12, 39–44, 59, 70–9, 117, 197
    disaster 12
    mitigation 70–9
    perception 7, 39, 40, 41–4
    reduction 20, 59, 117, 197
risk management 78
Royal Caribbean 105–9

safety 10, 15, 42–3, 53, 60, 70, 104, 109, 176
    community 43
    information 60, 72, 78
    personal 42
scale 27, 38, 45, 199
shopping tourism 149
social capital 9, 22, 25, 38, 40, 48, 156, 157, 159, 195, 196, 197; *see also* networks

importance 22, 38, 164
  and psychological resilience 161–3
social ecological system (SES) 154
social media 17, 55, 56, 59, 74, 126
social memory 22
social networks 21, 22, 38, 45, 48, 139, 146, 148; *see also* social capital
South Africa 146
South Korea 145
Sri Lanka 4, 85, 111
  Arugam Bay 85
stability 14, 22
  social 140
storm 41
storm-chasing tours 102
supply chain 150
sustainability 103, 193–4, 197
  and resilience 24–5
sustainable development 149
systems 9–13
  complex adaptive 21
  ecosystems 102
  physical 169
  tourism 3, 9–13, 40, 52, 199

Taiwan 14
taxation 106
taxis 106, 126, 127, 131, 132
Thailand 4, 15, 25, 84, 85, 93, 156
  Khao Lak 15
thanatourism *see* dark tourism
tourism
  demand 10, 13–14, 154
  and destination resilience 20–1, 25, 194
  experience 103, 190
  and individual and employee resilience 23, 154–9
  and organisational resilience 154–63, 198
  system 3, 9–13, 40, 52, 199
Tourism New Zealand 17
tourism system 3, 11, 40, 52, 194
  effects of earthquakes on 9–13, 199
transport 10, 108
Tripadvisor 78, 104, 105, 172
tsunami 4, 5, 15–16, 65–79, 85, 111, 173, 193
  2004 Boxing Day tsunami 4, 15–16, 22, 27
  risk mitigation 70–9
Turkmenistan 3

Twitter 17, 55, 56, 59, 60; *see also* social media
Tōhoku earthquake 4

Ukraine 174
  Pripyat 174
United Nations Economic and Social Council 109
United Nations Environment Programme (UNEP) 100
United Nations Office for Disaster Risk Reduction (UNISDR) 5, 37, 102
United Kingdom 107, 110, 143
United States of America 2
  California 2
  Los Angeles 2
  New Orleans 84, 85, 102
  New York 36, 84, 85
  Northridge 2
  San Fernando Valley 2
  Terlingua 174
  Ventura 2
United Nations World Tourism Organisation (UNWTO) 52, 54, 57
  Global Code of Ethics 103
urbanisation 52

Vietnam 93
  Sapa 93
volcanic activity 4
volunteer tourism 15
vulnerability 10, 19, 20, 36–8, 40, 42–3, 45, 46, 48, 85, 111
  defined 37–8
  destination 20, 21, 194
  disasters 42–3, 45, 46, 48, 70, 72
  infrastructure 194
  personal 159
  social 38–9, 194
  structural causes 111

Wallis and Futuna 3
water supply 139
Weichuan earthquake 14
wildfire 100, 156
workplace 156, 158, 159
World Bank 143
World Heritage 40

YouTube 56; *see also* social media

For Product Safety Concerns and Information please contact our EU Authorised Representative:

Easy Access System Europe

Mustamäe tee 50

10621 Tallinn

Estonia

gpsr.requests@easproject.com

www.ingramcontent.com/pod-product-compliance
Ingram Content Group UK Ltd.
Pitfield, Milton Keynes, MK11 3LW, UK
UKHW021943200326
4879IPUK00004B/72